The
Future

THE
FUTURE

Ronald D. Rotstein

A Lyle Stuart Book
Published by Carol Publishing Group

First Carol Publishing Group Edition 1990

Copyright © 1990 by Ronald D. Rotstein

A Lyle Stuart Book
Published by Carol Publishing Group

Editorial Offices
600 Madison Avenue
New York, NY 10022

Sales & Distribution Offices
120 Enterprise Avenue
Secaucus, NJ 07094

In Canada: Musson Book Company
A division of General Publishing Co. Limited
Don Mills, Ontario

Queries regarding rights and permissions
should be addressed to: Carol Publishing Group,
600 Madison Avenue, New York, NY 10022

Manufactured in the United States of America

10 9 8 7 6 5 4 3 2 1

Library of Congress Cataloging-in-Publication Data

Rotstein, Ronald D.
        The future / Ronald D. Rotstein.
              p.   cm.
        "A Lyle Stuart book."
        Includes index.
         ISBN 0-8184-0505-8 : $19.95
           1. Technological forecasting. 2. Twentieth century--Forecasts.
   I. Title.
   T174.R67   1990
   601'.12--dc20   89-7228
                                                              CIP

This book is dedicated to those who inspire true spirit and love through making the ultimate sacrifice, thinking of others first. These are the true angels of this world who will inspire mankind to reach into the heavens.

Contents

The
Future

Introduction

The Future isn't what it used to be.
—ARTHUR C. CLARKE

THROUGH TIME, the future was always somewhere out on the distant horizon. So far away that only prophets and mad scientists even dared to predict it.

People living in the sixteenth or seventeenth century were secure in the knowledge that the way they lived, the horse-drawn carriages that conveyed them, the tools they used, and the food they consumed were more or less the same as what their grandparents had had and their grandchildren would have.

Of course, there was always the threat of war, plague, and natural disaster, any one of which could bring upheaval and disruption. But kings came, kings went. Disease decimated villages and cities, and then became a bad memory. Spring always came.

Variations in the fabric of life were incorporated over lifetimes and generations, not a few years. Even the most historic and profound scientific breakthroughs rarely had immediate impact on how life and commerce were conducted. What, after all, was the Copernican theory to the butcher, the baker, the candlestick maker? The sun continued to rise in the east and set in the west, and what did it matter whether we revolved around it or it around us?

3

Then came the Industrial Revolution. With the steam engine and then a deluge of inventions tapping other sources of energy, such as electricity, the pace of change quickened. Yet even into the first half of the twentieth century the rate of change was manageable. It took a while for innovations to progress from idea to everyday life. The future still remained on the horizon.

The Wright Brothers took man aloft and that was wondrous; still, progress in flight and aircraft was spread over decades: the Jenny, the Spirit of St. Louis, Pan Am Clipper ships, turboprops, and finally jetliners. Mankind had time to adjust. Even manned flight to the moon was a ten-year adventure.

It took time for the telephone to spread from Alexander Graham Bell's lab to almost every household in America. It took decades for television, which was invented in the twenties, to broadcast Milton Berle hamming it up for millions of people.

Today that grace period for adjustment to development is disappearing. The distance between now and the future has been shortened dramatically by advanced telecommunications and computers. The world is interconnected as it has never been before. Scientists once learned of research and projects in other laboratories primarily—and belatedly—through results published in scientific journals. False trails were needlessly followed; work was unnecessarily duplicated. Now, scientific communities are interconnected by computer networks, fax machines, and conference phones. Information flows in minutes where once it took days or weeks.

Computers are enabling scientists to make quantum leaps in research. Within the next several decades there will be cures (not just treatments) for many diseases. Astronauts, mechanics, and submarine crews will be trained in increasingly sophisticated computer-simulated environments, where they will receive "hands-on" training without being in an actual spacecraft, or inside an engine, or under the sea.

Computers will transform our residences into "Smart Homes" programmed to keep us warm and secure, and to cost us less money. We will be entertained not by images on the flat screen of a television, but by three-dimensional holography that will make it seem as if Bill Cosby is teasing his television family right in our family rooms.

The future is now. It is in our laboratories, it is on our drawing boards. To see the future is only a matter of searching it out, finding the trends, exploring the implications. How will we live? What will we eat? How will we do business? And what must we do to be ready for what is on the horizon.

What follows then is about us—how we are changing the world, where we will be as technology and rapid scientific change affect our lives.

ONE

Computers: The Speed, the Power, and the Glory

IN THE BEGINNING it was the abacus, a simple board with rows of beads. Each bead denoted a number, each row a unit such as tens or hundreds. Push two beads up in the tens column and that meant 20. Push two in the ones column, four in the ten and one in the hundreds, and it showed 142.

The Babylonians probably invented the abacus, and it was such a useful extension of the human brain, giving man the ability to process more information quickly, that succeeding civilizations gratefully embraced it.

It is almost inconceivable that out of this modest beginning came the modern computer. Fifty years from now, personal computers with their keyboard and light pens will seem antiquated, quaint and fit only for museum shelves. Since the beginning of the computer era more than a century ago, there has been a steady and spectacular progression in the power and speed of these machines. Men have worked diligently in their efforts to create devices to relieve humans "of the work which has often times fatigued [their] spirit," as Blaise Pascal put it in the seventeenth century.

Hans Moravec, director of the Mobile Robot Laboratory of

7

Carnegie Mellon University, calculated this progression in his book *Mind Children.* "The amount of computational power that a dollar can purchase," he wrote, "has increased a thousandfold every two decades since the beginning of the century. In eighty years there has been a trillionfold decline in the cost of calculation. If this rate of improvement were to continue into the next century, the 10 teraops [ten trillion operations per second] required for a humanlike computer would be available in a $10 million supercomputer before 2010 and in a $1,000 personal computer by 2030."

Of course, we might wonder whether there is some point of saturation when the computer has gone as far as it can go. That is, of course, conceivable, but not in the near future. Why not? Because computer technology will continue to build and advance upon itself, just as the early computers such as ENIAC and Univac, which seem so feeble by today's standards, were used in building today's more advanced computers.

Speed is of the essence with computers—how many computations the machine can perform per second. Today's computers will operate as slowly as snails compared to tomorrow's. Special X-rays and the new "killer" technology of photonics will be the catalysts that speed up tomorrow's computers.

These special X-rays are produced by something called synchrotron radiation and are more powerful and concentrated than conventional X-rays. They will allow for incredibly close connections within integrated circuitry. With integrated circuits, hundreds of tiny transistors and other components are all placed on one tiny chip. The closer to one another the transistors and components, the faster the action.

Photonics has earned its "killer" nickname because it is expected to phase out electronics. Simply put, this technology transmits information—voice, video, or data—by pulses of light instead of electrons. That information can race along strands of glass thousands of times faster than electronic signals travel on copper wires. Early experiments in photonics allowed the equivalent of the entire Encyclopedia Brittanica to be transmitted in less than a second.

Photonic technology will lead to computers one thousand times more powerful than today's supercomputers. To put that into perspective, one of today's most powerful machines, ETA10, has reached nine billion calculations per second. These photonic computers, which are also referred to as optical computers, will be able to work so quickly because streams of photons can cross other streams of photons without losing data. As John Markoff explained in an article in the *New York Times,* "using light instead of electricity to perform calculations and to handle data is the computer equivalent of eliminating New York City's traffic jams by permitting cars and taxis to pass right through one another at intersections rather than collide."

Bell Laboratories unveiled an experimental optical computer in 1990, which it sees as the first step toward the next-wave of megacomputer. Alan Huang, who is in charge of the Bell research, characterized development of the prototype as akin to "the Wright Brothers. This is a technological milestone."

Bell Laboratories is expecting optical computers to be sold commercially by the next century.

Photonics and X-rays are only part of science's quest for the Holy Grail of speed. The architecture of the computer is also being redesigned.

Historically, computers work linearly. In other words, tasks are done sequentially on one processor. T. R. Reid, in his book *The Chip,* explains:

> A computer reduces every question . . . to the simplest possible terms: yes or no, one or zero, true or false. . . . An astrophysicist mapping the universe in the observatory needs to calculate the twenty-fourth root of arctan 245.6; to do it, he types the problem into his computer. The machine has to work through two dozen yes-or-no steps . . . just to figure out that someone has punched its keys. To determine which keys were pushed, and then to solve the problem,

will take another 100,000 steps, quite possibly more.

How fast the answer is reached depends on how fast the digital (linear) computer can take those yes-no steps. But what if the computer did not go from A to D in one direct line? What if it were able to divide up a task, go from A to B while simultaneously dealing with C to D? That is the basis of "parallel processing." A problem is broken into chunks and then divided up among several processors.

Some of today's largest supercomputers, such as ETA10, are designed with parallel processors and are being used to simulate and study complex, intricate phenomena such as thunderstorms, the changing population of insects over a quarter of a century, the fibrillation of the heart before an attack. These supercomputers are hyperfast. An Apple or an IBM PC would literally take centuries to do the same studies.

The more processors that can be linked, the faster the new supercomputers will be. The Japanese have said they developed a supercomputer with sixty-four parallel processors. Sandia National Laboratory in New Mexico has completed a project called the "hypercube massively parallel supercomputer" that has more than one thousand processors. Scientists there see applications for these supercomputers in weather prediction, tracking pollutants in the atmosphere or coolants circulating in nuclear power plants.

To test the power of the Sandia computer, the computer was programmed to determine the stress and strains a metal beam would experience under a certain amount of weight. The computer came up with the answers in only a week.

John L. Gustafson, a computer scientist working at Sandia, said of the project in 1988, "We have achieved results that most computer scientists thought impossible a couple of years ago."

Of course, supercomputers of this type will not go commercial until their costs drop. At the end of the eighties they cost as much as $25 million. It was, therefore, of great interest when General Microelectronics Inc., a contractor for NASA's Ames Research Center in California, announced in 1989 that they had developed a computer with eight processors (or nodes) for a mere $300,000. Furthermore, on one test it was able to perform at 70 percent of the speed of one of the Cray supercomputers. NASA hopes to use the new computers to determine how spacecraft are affected by various forces.

It is ironic, with parallel processors on the brink of commercialization, that the Japanese had never intended to get into that field. In 1982, they announced, with great fanfare, a ten-year, multibillion dollar government-backed campaign to develop what they called the fifth-generation computer. The Japanese saw this as their chance to get ahead of the United States in computer technology. Parallel processing had not been the route they had originally planned. No, they had their sights on a completely different type of architecture, one that attempts to imitate the abstrusely complex operation of the human brain—neural net computers.

To understand the concept of neural nets, a short course in what we know of how the brain works is required. What we are concerned with here are neurons in its web, cells within the brain that carry information. Neurons are made up of a cell body and branch-like structures called dendrites, which receive information from other neurons. Also attached to the cell body is a tubular fiber called the axon, which has its own branches. The axon carries information away to other neurons. There is an infinitesimally small space between neurons, called a synapse. The electrical impulses of information must jump this gap.

There are between 100 and 200 billion neurons in the human brain, each connected in dendrite webs to a thousand or more other neurons. When a word is uttered, the sound waves reach the brain as electrical signals. Millions of neurons react, each sending out signals to a thousand or so other neurons. In this way, millions of neurons are processing information simultaneously. There is much that is not understood about the workings of the brain, but it is believed that all functions of the brain are produced by this interaction between the neurons.

This arrangement produces the human brain's complexity and

subtlety. The linear computer is adept at rapidly calculating a rocket's trajectory or finding the square root of a number to the ten-thousandth place. But ask it to differentiate between a rose and a tulip or to find a path through the toys on an eight-year-old's floor, and it has a problem. That is because the digital computer is not designed to fill in gaps, to recognize similarities, to take leaps of faith based on past experience. It cannot perform reasoning tasks a toddler still in diapers does with ease.

What scientists are searching for are computers designed to think like humans, that can learn from example and pattern recognition. Scientists working on this type of computer architecture are trying to duplicate the brain's structure with electronic components. This technology is in its very early stages, and there are many naysayers who doubt it will go anywhere. Tomaso Poggio at MIT is one of those naysayers. In 1989 he said, "I have yet to see any application that I find new or different, or where they [neural net computers] are more powerful than existing computer systems."

Yet despite such doubt, progress is being made. At Johns Hopkins University, Nettalk has been developed. This computer can teach itself to pronounce words by listening to a tape recording of a child talking. At first the computer talks gibberish then after a few hours of repetition, the words become intelligible until, after twelve hours, it has a vocabulary of one thousand words. "It surprised us. We didn't expect it to do this well," said one of its developers, Terrence Sejnowski.

Within the next twenty years tremendous advances will be made with this technology, although even then neural net computers will still be rudimentary in comparison to the human brain.

As MIT's Poggio has pointed out, "Man has spent three thousand years thinking about what is thought. We shouldn't expect artificial intelligence or neural nets to solve it in five, ten, or even one hundred years."

So far this has been a discussion of computer hardware, but on the horizon are great changes in the software as well, changes that will have a major impact on even the most casual user. For what scientists are wrestling with now are ways for humans and machines to interact—or interface—more easily.

We have already come a long way from those early dinosaur-sized computers that spewed out tons of punched cards. The introduction of the mouse was an enormous leap toward easier access to computers. It made computers "user-friendly." You could get the computer to perform by moving the mouse rather than typing in precise instructions on the command line and praying you did it correctly. But the mouse is only a rudimentary beginning.

Humans do not naturally communicate by moving a device around on a tabletop. We speak, listen, watch, gesticulate. The goal is to produce communications with computers that will be similar to the give and take of communications between humans. As John J. Anderson wrote in the March, 1989, issue of *Macuser,* "the ultimate goal of computer technology is, in a sense, to make the computer disappear, that the technology should be so transparent, so invisible to the user, that for practical purposes the computer does not exist. . . . In its perfect form, the human interface is entirely that—it allows a person to interact with the computer just as if the computer were itself a human."

The human user of the future will not be distracted by the computer. The user will need only be involved with the subject at hand. Interaction with computers will become simultaneous with usage.

One of the most important elements in achieving a more facile interaction between man and machine will be natural speech recognition. The computer will have to recognize verbal commands. This is extremely complicated since people have different accents, intonations, and, regrettably, a tendency toward sloppy speech patterns. We elide, run our words together, mumble, all major barriers for a machine trying to make any sense out of what we are saying. Also, we tend to speak in a natural language that takes great liberties with grammar and the meaning of words. Yet it is predicted that the computers of the future will be able to decipher our fuzziness and respond appropriately.

The computers' responses will be verbal with the capability of interrupting us for clarification. At some point we will be able to control what is on the screen by a wave of the hand or the movement of our eyes. The computer will have built-in tracking devices sensitive to such movement.

Now if all this sounds far-fetched, ask to borrow a Nintendo Power Glove from a ten-year-old. The video game industry has seen the future and is already marketing it. The glove is equipped with sensors and fiber optic cables so that when the ten-year-old motions at the screen for Mario to move right, Mario moves right. Jump, shoot, duck, all the things the child usually uses the control pad for, he can accomplish by moving his hand.

This kind of technology is moving us into the sphere of computer-controlled virtual reality.

Imagine this: You are Mister Blandings and you wish to build your dream house. Rather than trying to visualize the finished product from a sheath of architectural drawings, you put on what has been dubbed your "magic wardrobe," consisting of helmet, gloves, and body suit. You will then be able to enter a simulated computer-generated environment where you can interact with what is around you. You will be standing in your dream house, the four walls surrounding you. If you do not like the placement of the windows or doors, with a wave of your hand they can be moved or their size and shapes changed. You will be able to go from room to room with the sense of actually being there.

As Jaron Lanier, a founder of VPL Research, a pioneer in the virtual reality movement, said, "It's a new level of reality. There's never been another one except for the physical world, unless you believe in psychic phenomena."

The applications of virtual reality range from frivolous to profound. Two people continents apart will be able to play tennis with each other. Chemistry students will have the capability of moving among molecules. Astronauts, pilots, and mechanics will all receive hands-on simulated training.

These gloves and helmets already exist. But there are several problems that have to be solved. One is that computer-generated images are not yet lifelike. It takes several hours of programming to get one lifelike frame. To create a virtual reality setting, thirty frames per second would be needed. It will take the more powerful computers of tomorrow to accomplish that.

Another impediment is feedback. The user's real hand has to feel what the simulated hand has grasped or touched. Researchers also worry about health effects. Will the user get so disoriented he experiences nausea like the less interactive flight simulators of today can produce?

These problems are not insurmountable, and many researchers expect the day is not far off when virtual reality will be a reality.

After all the years of prediction and promises of robotic housecleaners, streetcleaners, and farm workers, the day is near when such devices will exist and be accessible.

The robots of today are relatively simple-task devices. Put them on the car assembly line and they weld, paint and insert pieces. Others have a minimal vision system which they use to locate specific pieces on a conveyor belt. When such systems have been coupled with robotic arms, they can be used to assemble calculators, typewriters, and other such small items. To overcome lack of mobility some factories have installed sensor pathways in factory floors so that robots can follow specific job routes.

But a robot that can vacuum your house? That requires a much more sophisticated device. It would be useless to prescribe a fixed pathway. Bookbags get left on the floor. The basement door gets left open. These are impediments and potential dangers to a robotic housecleaner. Scientists are predicting that robots will be programmed to deal with such obstacles as computers take on general learning abilities. The machines of the future will come equipped with conditioning software that will make the robots respond much the way Pavlov's dogs did. Positive and negative messages will accumulate in the computer's catalog. As situations arise, the computer will run through its catalog to ascertain what kind of reaction the obstacle had evoked in the past.

However, these scaled-to-life robots would not elicit the most wonderment from any errant time travelers from the past. That would fall to something far more fantastic, the nanobot.

Nanotechnology is miniaturization to microscopic size. The prefix *nano* stems from Greek, meaning "dwarf." When scientists use it, nano refers to a billionth. Nanotechnology, therefore, is working on a scale of one-billionth of a meter.

Scientists in this field believe that someday we will be able to manipulate individual atoms, assembling them as we wish into what we wish. As K. Eric Drexler, of MIT's artificial intelligence lab, has explained, "The difference between coal and diamond, or between cancerous and healthy tissue is purely a matter of how the atoms are arranged."

There is talk of someday having armies of nanobots being sent through a carburetor to clean it from the inside or injected into the bloodstream to eat through cholesterol deposits or to eradicate viruses.

Furthermore, according to Drexler, nanotechnology will create computers that "are a trillion times faster than ones we have today and a million times more energy efficient, as well as machines small enough to repair living cells, and much more."

Scientists at Bell Laboratories have already manufactured a turbine that is only .005 inch in diameter. That is no larger than the decimal point on this page. "You have to be careful when handling these things," one of the Bell Laboratories scientists has warned. "I've accidentally inhaled a few right into my lungs."

Tiny pressure sensors—micro-sized, not nano—are being used on the space shuttle *Discovery* to measure hydraulic and cabin pressures. It is estimated that by 1995 the typical car will be equipped with fifty different microsensors that will be used to release airbags, control antilock brakes, and survey other engine functions.

While nanoassemblers seem to be taking computer technology to the outer limits, they are no less fantastic than some of the visions

of Hans Morevac. Morevac makes a distinction between the brain—
a corporeal entity—and the mind. He speaks of the possibility of one
day transferring your mind, with all its knowledge and memory, into
a machine. He suggests several possible methods, from a robotic
surgeon scanning your brain and transferring the measurements to a
computer, to an implanted computer picking up and recording the
messages sent from one hemisphere of your brain to the other until
it has a complete model of your mind.

Surreal? Out of the question? Beyond the realm of possibility?
Such conjectures, more than anything, illustrate that with the quick-
ening pace of computer technology, the boundaries of the realm of
possibility are no longer within sight.

TWO

Don't Forget to Turn Out the Lights

As SOMEONE pulls into Meryl Streep's driveway, her doorbell rings, giving the actress time to fix her hair and adjust her demeanor before the visitor even reaches the front door. Whether she knows it or not, Streep is a pioneer. She is in the forefront of the computerized home—the "Smart House."

While in the previous chapter we talked of speed and power, scientific and medical applications, the impact of the computer is invading our homes. This is not about keeping the household budget on your PC or going on-line to services such as Prodigy, where through a modem you can shop at your neighborhood grocery store, get toys delivered from J. C. Penney's, or book a seat to Chicago on the American Airlines reservations system.

We are talking about computers running your house, from the garage door opener to the hall light. Using remote control and microchip technology, new houses will be built that are completely automated.

A centralized computer will be able to monitor the heating and cooling systems to keep them at peak performance. The computer systems will direct the water heater to fire up in the early morning

19

hours when electricity rates are down, then turn on the dishwasher; follow the progress of a child on her way to the kitchen for a drink of water, turning on the lights ahead of her, turning off the ones behind; show on screen the movements of everyone and everything in and around the house. You will be able to program your bathtub to have hot water waiting when you wake up. Or the lights to dim as your baby falls asleep.

For this to work, computer chips will be built into appliances and a new type of wiring system will be utilized. Conventional homes today have as many as five systems—120 volt lines for lights and small appliances, 240 volts for heavy-duty appliances, and separate wiring for telephone, cable television, and security.

At the heart of the Smart House is a single electrical cable that combines all the functions of the various wiring of today. The new wiring will mean that any device, be it phone, VCR, computer, stereo speakers, or refrigerator, can be plugged into any outlet. Furthermore, no outlet is live until something is plugged into it. Not only will this wiring provide savings in electrical costs, but parents will not have to worry about baby getting a shock when curiosity leads him to stick his finger into a socket.

David MacFadyen, president of the National Research Center of the National Association of Home Builders, has been a long-time advocate of the Smart House concept. "I think we're going to see American housing increasing in its functionality very dramatically for the next twenty years," he said in 1988. "This will increase the pressure for the replacement of houses. In the United States, we see houses going out of stock at the rate of about 2 percent a year.

"The more we can make new housing dramatically better for the occupants, the more we will see new housing competing success-fully against the existing housing stock in the marketplace, and we will see more of the existing housing stock becoming obsolete. Homes will have to be upgraded dramatically by remodelers or replaced with new construction."

Homeowners are already finding they have to upgrade their elec-tricity if they are going to be able to outfit their homes with all the current gadgets. The need to do so will only increase as more homes get more computers, compact disc players, advanced versions of food processors, and all the other accoutrements of modern living.

Homeowners of tomorrow will want to be ready to plug in the entertainment centers of the future, which will one day include full-motion holography, a computer technology that creates three-dimensional images that actually appear to be in the room with you. You will not only be able to sit and watch an actor perform right in front of you, you will be able to walk around him and watch the action from behind.

THREE

Business Not as Usual: The Third Industrial Revolution

IN THE LAST couple of decades, business forecasting and prognosticating has mushroomed into a major enterprise in the United States, with hundreds of thousands of players. At times it seems as if each has his or her firmly held, loudly expounded, and totally divergent view of what happened to the American economy, where it stands now, and where it is going. These business prophesiers range from the gloomiest pessimists to the sunniest of Pollyannas. As New York financial adviser Sonja Kohn once said, "If you laid all the economists in the world end to end, they still couldn't reach a conclusion."

Of course, the doom proclaimers had much to point to in the seventies and eighties. Factories were shamefully out of date. Too much money was spent on gadgets and toys, too little on education. The American worker was viewed as lacking a work ethic. American companies seemed to be languishing on the home market and losing their competitive edge. Industry after industry fell to the onslaught from the Far East.

We have progressed from lender nation to borrower nation in a soberingly short period of time. Our personal savings rate, which hit an all-time low in 1987, was only a third of Japan's.

It was indeed a good time for the predicters of panic and cataclysmic depression such as Ravi Batra, whose best counsel was to put everything in cash and hide it.

Still, the American economy, as we approach the next century, does not appear to be on the brink of dissolution. True, it is unlikely we will ever regain the position of dominance we held after World War II, but as Charles R. Morris wrote in the September 1989 issue of *Atlantic* magazine, "That role is gone forever, and no one would wish it otherwise. The income of urban adults in the major industrialized countries is now practically uniform. That was the explicit objective of American statesmanship at the end of the war, the crowning adornment of our post-war foreign policy, a grand aim expressly adopted, pointedly pursued, and unambiguously attained."

If that is so, the real question becomes: Where do we go from here? If industrialized urban incomes and economies are in parity, what does that mean for America? Clearly, on an industrialized and technological level, there will be a new set of rules for playing the game.

From all indications, there will be great changes. Changes in where we do business and how. In which industries and areas will grow. In which jobs will be in demand.

In many respects the seventies and eighties were sobering, cold-water-in-the-face decades. However difficult and costly, they were cathartic. American businesses had to face the unpleasant fact that there was no guaranteed lock on any market, that there was no such thing as a sure sale.

The figures told the story. In 1975, American companies supplied the world with machine tools. Ten years later the United States had scarcely any machine tool exports. The American share of semiconductor sales went from 60 percent to 40 percent worldwide. Within ten years, imports to the United States doubled, and by 1986, 70 percent of American products were competing directly with foreign-made goods. Even IBM, so stalwartly American, was importing 37 percent of its parts.

The figures and statistics were indicative of a trend that will only become more pervasive in the coming years. It was a trend that, according to the United States Department of Commerce, many American companies were slow to recognize and upon which they were slow to act. The trend was the globalization of the world marketplace.

While Marshall McLuhan may have been referring to communications when he spoke of the global village, the marketplace has developed globally as well.

This concept covers more than exporting. It encompasses joint ventures, partnerships, licensing agreements, and overseas plants and marketing units.

The American auto industry, so severely hurt by foreign competition, has already faced the necessity of global strategies.

A "Made in the USA" bumper sticker displayed on a Plymouth Voyager may make the driver feel patriotic amid the Toyota and Nissan owners surrounding him on the highway. But made in the USA is a declaration with little meaning now, and it will have even less as we approach the next century.

By the end of the eighties, Chrylser had the lowest percentage of American-made parts in its vehicles of any Detroit automaker. Furthermore, it owned 24 percent of Mitsubishi Motors, which in turn owned part of the Korean company Hyundai. Mitsubishi and Chrysler also have a joint-venture factory in Normal, Illinois. Cars made there will be sold under both nameplates.

It is very much the same with the other American automakers: Ford owns part of Mazda and has joint ventures with Nissan and Volkswagen. General Motors is buying into Saab and owns 41.6 percent of Isuzu, which in turn is joining forces with Subaru. And so it goes.

Global joint ventures will continue to proliferate throughout the nineties.

Going global has many advantages. One is not having to rely on local suppliers. Take the case of the Will-Burt Company of Ohio, a maker of truck parts and mobile-radio transmission towers. Will-Burt was a healthy, thriving company until the mid-1980s. Then it lost a large contract with Caterpillar, Inc., which found it could get parts cheaper from Belgium and Brazil.

Faced with the loss of a large portion of its business, Will-Burt did some reassessing. It found it could not match the foreign prices because of the rise in the value of the dollar. So Will-Burt started getting casings from China and welded parts from Indonesia. It not only got back the Caterpillar contract, but won a new one with Mack Truck and is in the running for Volvo business, as well. Will-Burt had always been a 100-percent American concern. But as its chief executive officer Harry E. Featherstone explained, "We realized . . . that we were operating in a worldwide market."

Another step in the globalization process is overseas production. As Michael Plumley, CEO of Plumley companies, a Texas family-owned auto-parts maker, sees it, "You can't sell very much abroad without switching to production overseas." Plumley had tried direct export to Europe but gave up after finding the couplings or fittings needed there were different in each country. So now it ships hoses to a European partner, which then attaches the appropriate fittings for the customer. "We don't have the money for our own European factory or distributorship," Plumley explained, "but ten years down the road, that's the way we'll have to go."

No matter where the factory is located, production methods stand to be changed as the marketplace expands.

Again, Charles R. Morris in *Atlantic*: "The drive for the global market share is forcing what Stanley Feldman, of Data Resources, calls the Third Industrial Revolution. A wholesale reordering of production technology is under way—computerized factory schedules and inventory to cut costs, intelligent production machinery that can shift processes in the middle of an assembly line. The object is to produce local products adapted to local markets, but reap the world economies of scale in research and development, raw materials, sourcing, and production balancing."

Industry after industry is recognizing the need for globalization or is falling under its influence. Western Electric, the manufacturing arm of AT&T, makes telephones in China. Whirlpool washing machines will be made in six Common Market factories as part of a joint venture with the Dutch-based giant, Philips. Texas Instruments has joined forces with a Taiwanese company to produce chips in Taiwan.

Hollywood is another industry following the trend. Sony's $3.4 billion buyout of Columbia Pictures in 1989 was just one more sign of the times. The Japanese had come to dominate the "hardware" of the entertainment industry—televisions, videocassette recorders, and such—so their next move was to go after the "software."

As Sharon Patrick, head of the media sector of consulting firm McKinsey & Company, explained at the time, "If you want to get into position in the feature film industry, you have to own U.S. The market is here. The talent is here. The cultural product travels better."

Time, Inc., Chairman J. Richard Munro also predicted that "by the mid-1990s, the media and entertainment industry will be composed of a limited number of global giants."

The figures would bear him out. In 1988 one half of the $66.1 billion in revenues of the one hundred biggest media companies was generated by the ten largest firms.

The globalization movement will be further intensified by Europe 1992—named after the target year for removing internal trade barriers between the European Economic Community nations.

The twelve EEC countries—Belgium, Luxembourg, the Netherlands, France, West Germany, Italy, Great Britain, Ireland, Denmark, Greece, Spain, and Portugal—are planning to establish a free-trade area with no tariffs, custom duties, restrictive regulations, or visas. Europe 1992 is seeking to conform national policies and level taxes.

As it stands now, there are twelve different sets of rules, regulations, and forms to fill out. In 1989, the same model BMW, because of different tax structures, cost $17,000 in West Germany and

$50,000 in Greece. From country to country there are differently sized electrical outlets and plugs; they will be standardized. Even little bothersome questions, such as how to classify snails, will be worked out. (Currently France lists snails as fish, while the British put them under "small livestock.")

Some steps have already been taken toward 1992. Now truckers can cross borders with one piece of paper instead of stacks of forms. This is the first time that has been possible since Napolean enforced a modicum of uniformity on Europe with the assistance of his armies.

Europe 1992 will mean an interconnected economy of 320 million people with a 1987 gross national product of $4.3 trillion. The United States population in 1987 was 244 million and its GNP $4.5 trillion (more evidence that our postwar goal of achieving parity among the industrialized countries was reached).

The magnitude and potential that such a market offers is enticing Americans overseas, further fueling the trend towards globalization. A few hundred American law firms are opening offices in Europe in anticipation of increased business when the trade barriers come down. Trammell Crow Company, America's largest developer, is quickly building offices in Europe, where there is only a 3 percent vacancy rate, to accommodate the expected influx of American concerns.

General Motors chairman Roger B. Smith has said, "We think economic integration strengthens Europe and creates opportunities for U.S. companies—provided, of course, the EC-wide standards don't discriminate against American products or firms."

Smith thinks Europe 1992 will succeed. "I believe the tale of [Europe 1992] will end with a dynamic, unified European market [and] a new vigorous player on the world economic stage."

To take advantage of growing markets, whether in Europe or on the Pacific Rim, the United States will need the technology to produce competitive products. There was a time when America was secure in its technological lead. Our free-market, competitive approach seemed to work well. Companies had their own research staffs which labored in secret until a product or process was per-fected and ready to hit the market.

For projects that might entail more pure science, the nation was able to rely on the Defense Department and for many years the Bell Laboratories. With the Bell System enjoying the cushion of the AT&T monopoly of telephonic communications, its research "House of Wonders" could wander into fields of study without always hav-ing to worry about the balance sheet. Out of Bell Labs came the first digital computer, the transistor, the first solar battery, lasers, fiber optics, and much more.

Then came the fight over divestiture and the eventual loss of monopoly standing. The concern then arose that what had been called "the greatest research laboratory the world has ever known" would become more limited in scope. The new AT&T might not be as willing to spend money on projects with no direct bottom-line application.

At about the same time as divestiture, other countries were put-ting research and development on their national agenda. But Japan, for one, did not follow our free-market approach to technological innovation. Instead, it tried a more centralized and structured way of encouraging research, using as a vehicle its Ministry of International Trade and Industry (MITI). At one time it was suspected that MITI's primary function was to examine United States technology and copy it, and it may well have been. At the end of the eighties, however, MITI had a decidedly different function.

As part of its program, every two years the ministry asks Japanese industrial leaders and scientists to list what they believe are the most important technologies. The lists have ranged from genetic engineering to supercomputer development. The Japanese govern-ment then earmarks these technologies for priority treatment in the form of tax breaks.

When Europe tried to close its technological gap, it more closely echoed the MITI example than ours.

Back in the eighties, the Common Market set about transforming old-line industries with its "Shared Technology Program." To cure

Eurosclerosis, high technology is being used to change antiquated nineteenth-century smokestack industries into twenty-first-century models of productivity. As part of this effort, the Common Market in 1985 put $166 million into its BRITE (Basic Research in Industrial Technologies in Europe) program to fund research in a wide variety of fields with applications throughout Europe.

There have been numerous successes resulting from the cooperative endeavors. Superadhesives invented in Brussels are being used by Spanish toy manufacturers. Enzymes separated in British medical centers are used by the Swiss to manufacture biotechnology products. Industrial polymers developed in the Netherlands are being used as specialty coatings in Luxembourg.

Whole industries are being converted and revived, none more dramatically than textiles and clothing. From 1975 to 1985, under the onslaught of inexpensive Far Eastern imports, European employment in these fields plummeted 26 percent. The garment workers who survived the massacre spent 90 percent of their time twisting fabric through rusty sewing machines.

Technology-sharing brought dramatic changes. Computer-driven factories have freed workers from labor-intensive sewing and brought down the cost of production. It is believed that within the next decade, almost all European fabric factories will have computer-controlled production.

The idea that advanced technology is essential for survival has even reached into the haute couture houses of Paris, where microcomputer systems are being used in fabric design.

Another project, Eureka, followed BRITE. It was created not only to fund research but also to develop cross-frontier business ventures using the research findings.

From Eureka came a joint venture between Marconi Electronics Devices, a subsidiary of Britain's General Electric, and the French electronics company, Thompson. They are developing new electronic switching devices for the next generation of computers, which will use photonic lights.

This cooperative effort between foreign governments and industry has not gone completely unnoticed in the United States. Many analysts have begun to fear that individual American companies do not have the resources to compete on a global basis with Japan and the Common Market companies.

There are many in American industry and government who have called for, at the very least, a loosening of antitrust laws in order to give U.S. companies a fighting chance worldwide.

Pat Choate, a vice president at the high-tech conglomerate TRW, has pointed out that "Every other industrial nation . . . recognizes that there are certain industries—HDTV [high definition television], semiconductors—there are 'linchpin' industries."

The view is that these linchpin industries, for the health of the entire economy, should receive some sort of help or preferential treatment from Washington.

The United States is being forced to accept the reality that we must find new ways of financing technological development. Even Ronald Reagan's most laissez-faire of administrations accepted that a new approach was needed—at least in the semiconductor industry.

And little wonder. In 1979, American companies held 55 percent of the world market for semiconductors, the tiny chips used in electronic equipment. By 1988, that share had dropped to 37 percent, with Japan accounting for half of the semiconductors sold worldwide. Furthermore, with the help of those tax breaks, Japanese companies by 1989 were spending 50 percent more on research than American companies.

These figures had the Pentagon worried. Advanced weapon systems depend on semiconductors. No matter how friendly the supplier nation, the Joint Chiefs of Staff had no desire to become dependent on a foreign country for such a crucial part of our defense system.

Out of this worry came Sematech.

Sematech is a consortium of top American computer and computer chip companies, including IBM, that receives government funding. It started out with $200 million from the fourteen companies involved and the federal government. Its purpose is to keep the United States in the semiconductor business by developing advanced chipmaking technology and getting ahead of Japan in research.

In some respects Sematech is an experiment in what Commerce Secretary Robert A. Mosbacher and U.S. Trade representative Carla A. Hill call "industry-led policy." Its goal is to enhance innovation through joint research efforts.

A 1989 Commerce Department report gave the effort high marks, although it did note that other industries without the same clear-cut, long-term research goals might not be suited for similar consortiums.

Despite apparent early success, the idea of government-backed research consortiums in general and Sematech in particular have not been without critics. Budget Director Richard Darman, for one, is not happy with the government trying to "pick winners and losers." It was rumored that it was Darman who was behind reports in December 1989 that the Bush administration was going to cut all funds to Sematech.

As it turned out, the administration backtracked quickly, in part due to a National Advisory Committee on Semiconductors report as well as one from the Economic Policy Institute. Both concluded that the government had to take a more active role in American high-technology industries.

Ian Ross, president of Bell Laboratories and chairman of the National Advisory Committee, stated at that time: "Every trend you look at is in the wrong direction in the United States."

Congressman Norman Mineta (D-California) put it less kindly. "The Bush Administration appears content to allow American high technology to wither away. It is as though they woke up one morning and decided to throw away our future."

Whether Sematech survives remains to be seen. But, unquestionably, the United States needs new strategies, and big-scale joint ventures are the key to at least some future technologies for a global market.

While we are searching for those strategies, it should be kept in mind that globalization has only been possible because improved telecommunications and the computer have made the globe smaller and more manageable.

Thirty-five years ago it might have taken several days for an air-mail letter to get to Europe and several hours to get a call through. By the end of the eighties, a letter to England could be faxed in a matter of minutes and the speed of a call depended on how fast your fingers could hit the buttons on the phone.

Thirty-five years ago the amount of data and information that could be processed was limited by the number—and intelligence—of a company's employees and the amount of available file-cabinet space.

Despite coming a long way, information technology—as the scientific journals call it—has not come close to cresting. Detmar W. Straub Jr. and James C. Wetherbe, management professors at the University of Minnesota, see the 1990s as a time when refinement of existing technologies and the introduction of innovative ones will "revolutionize the way people work and process information."

Information transformed into knowledge will be the key to success in the 1990s. Raw data floating around will not help businesses make the fast short-term decisions they will need to compete, nor will undigested facts tucked into a computer's memory aid in the long-range planning that will be necessary to move profitably into the next century.

Computers, although ubiquitous, are actually greatly underused. In many operations there has been a dichotomization—the computer people and everyone else. What the 1990s will see will be a "demo-cratization of the accessibility of information [within] organiza-tions," as Mike Hammer, president of Hammer & Company, put it. Information will open up to everyone, not just to those who know what to type on the command line of a computer.

Overthrowing the dominance of the keyboard is believed to be a key factor in opening up the knowledge bottleneck. Many people just do not like computers. Typing on keyboards is not how they are used to communicating. They talk. They listen. They respond. And that is the direction in which computers are moving. It is predicted that by the late 1990s computers will be able not only to receive verbal commands, but will also be able to interact verbally. Instead of having to type "call 1989 XYZ Company billing" on the command line, the human will be able to say "1989 XYZ Company billing" and the computer might respond, "XYZ of Denver, Colorado, or XYZ of Peoria, Illinois?" and then come up with whichever information was actually desired.

While that natural kind of interaction technology is being perfected, there will be interim devices to soothe the user and make the systems more accessible. One, of course, is the mouse. Screens that respond to touch will also reach the marketplace.

While the 1990s will see easier access to information, it will also see faster dispersal of information. Raw data has to be put to use to become knowledge. Knowledge in the age of globalization must be disseminated rapidly and efficiently over great distances, necessitating more sophisticated communications technologies. Companies with offices in New York and Melbourne need systems that will allow workers to communicate as if they were separated by a wall and not some ten time zones.

Tremendous demand for the fast flow of information is evidenced by the so-called fax revolution. In 1985 if you'd said fax to someone, they would have looked at you blankly or thought you were mispronouncing the word "facts." By the end of the eighties, if you did not have your own facsimile machine, you could run to the local drugstore to "fax" a document. (A sure sign something has been totally incorporated into the American cultural fiber is when a noun becomes a verb: "I'll Xerox the article and then fax it to you.")

But for all their charm, fax machines are merely a low-tech bridge to what will come in the 1990s. Though off to a slow start, E-mail (or electronic mail) is expected to achieve far greater acceptance and eventually supplant fax.

E-mail allows you to send information directly from your computer to someone else's computer. It eliminates what is called telephone tag—he's not in, then you're not in, etc. And it eliminates paper. Offices of the future will be increasingly paperless. With fax, you first have to create a document, send it, and then the receiver, if he wants to store the information, puts it into a file cabinet or transfers it to computer. E-mail starts in the computer and ends in the computer with no time-consuming intervening steps.

And perhaps more important, users of E-mail have found that it enhances the "democratization" of business information. A stock room manager in the Boise warehouse might hesitate to pick up the phone and call the company president in New York. But being able to zip the message cross country, screen to screen, is less forbidding. Studies have shown E-mail encourages workers at all levels in a company to communicate more freely than telephone or paper mail. (U.S. Post Office and even Federal Express letters are referred to as "snail mail" by many computer users.) There is a built-in informality to it, almost like bumping into the company president four times a day at the water cooler. That promotes casual interaction—which generally leads to the transfer of information.

In one respect, E-mail's early history was similar to that of the telephone system. In the telephone's infancy, there was no monopoly on local companies. Someone who lived on East 88th Street in New York City might not have been able to call someone on West 88th Street because they were serviced by different phone companies that were not interconnected. So it was with E-mail. Western Union did not connect with AT&T Mail which did not connect with MCI Mail which did not connect with U.S. Sprint's Telemail. By 1989, the various providers saw the futility and nonprofitability in this proprietorship and opened up the electronic gateways, making it much easier to send messages.

But the sending of messages is only the first step in what E-mail has to offer. Programs are being developed to expand electronic mail. For example, the Lotus Development Corporation, of Cambridge, Massachusetts, has introduced a program called "Notes." In essence, Notes takes the incoming messages and sorts them according to categories. Notes is an example of "groupware," in which specified pieces of information get forwarded to specified groups of people. For example, only E-mail of interest to the advertising department of Revlon would be directed to that department.

According to Brownell Chalstrom, director of telecommunications and networking at Lotus, "If you look at traditional electronic mail it really falls short. People typically use mail for short things. We offer more structured communications."

E-mail will be opened up further with what is called "unified-messaging architectures." IBM, Digital Equipment Corporation, and AT&T are all trying to be the first with this design that would allow the transmission and receipt of messages using video, speech, graphics, and text.

Another communications technology expected to flourish as it is refined in the coming years is teleconferencing. Teleconferencing is nothing more than one conference held simultaneously at several different sites, with the participants linked by satellite. The least sophisticated of these systems would be an audio link-up and still picture of whomever is speaking. More advanced teleconferencing includes complete visual transmission via satellite, so that the lawyers in Topeka can see on a screen the lawyers in New York as they speak and move.

Having access to this technology would save time and money spent on travel and involve more personnel in decision-making. It will also reduce the wear and tear on people who would otherwise be constantly on the road.

Teleconferencing did not get off to an auspicious start. The old Bell System introduced something it called Picturephone, a "video conferencing phone booth," at the end of the seventies. These were public meeting rooms in various cities connected by satellite. By the mid-1980s, AT&T had begun closing many of the facilities.

One reason for the failure of Picturephone was inherent in its design. The advantage of teleconferencing is supposed to be its convenience. That advantage was dissipated by having to travel to the conference site. "People just don't want to get into their cars and drive off to some public phone place somewhere away from the office," explained Phillip L. Gantt, of Peirce-Phelps, Inc., a company involved in designing and installing teleconferencing rooms.

Large, far-flung companies are interested in their own teleconferencing system. As these systems come down in price because of advances in technology—such as the use of direct broadcast satellites that beam video signals to relatively inexpensive rooftop satellite dishes—more corporations will begin using this technology. Irving Goldstein, president of Communications Satellite Corporation, has predicted that by the year 2000 "it will be impossible to conduct business without [video teleconferencing] capability, just as you can't conduct one without a telephone today."

Arno Penzias, the 1978 Nobel Prize winner who is director of research at Bell Laboratories, sees another advancement that will greatly improve teleconferencing capabilities—Integrated Services Digital Network, or ISDN. With ISDN, different types of information in different forms such as voice, data, graphics, pictures, can be transmitted and received simultaneously. ISDN could give an executive attending a teleconference access to all his files. He could draw from the files and transmit them to other participants, somewhat like passing a sales graph across a table.

But teleconferencing, just like across-the-table conferences, is of a formal nature. There is also a need for the informal, ongoing contact between far-flung sites. Xerox began working on such a project in the mid-eighties with its linked-office experiment.

"There's more to working with people than just sitting down and having meetings," George Goodwin, a cognitive psychologist with Xerox explained. There's also the need to be able "to walk up to someone's office, lean on the door and say something random."

Using existing technology, the project hooked up a lab in Palo Alto, California, with one in Portland, Oregon, using speaker phones, television cameras, big-screen video monitors that were constantly on in common areas. A worker could wander into a common area in Palo Alto, ask for help on a technical problem, and get an immediate answer from someone in Portland who could be seen and heard on the video monitor.

One problem the experiment ran up against was keeping a record of what went on. Just making a video tape would be next to useless if selected images could not be called up when needed. Otherwise someone would have to wade through the tape until finding the section he needed. Again, data in raw form is of little value. And that will be another technological trend of the future—the development of more sophisticated and readily available methods of storing, sorting, and retrieving information.

As Arno Penzias of Bell Labs sees it, "There are a trillion lost pieces of paper in this country stored away in filing cabinets. We are being drowned in information. We need to find a way out of this morass."

Bell Labs is working on one device to alleviate the problem. It is, essentially, an electronic filing cabinet that would combine features of a personal computer, a copier and a fax machine. As information arrives in an office, whether in the form of letter, faxes, E-mail or even verbally—the "filing cabinet" would organize the information and store it. It will even let the user know when important information has been received. It is expected this machine will be on the market sometime in the nineties.

Companies or operations that need far greater storage capacity— such as the United States Census Bureau or the Internal Revenue Service—will make greater use of optical storage technologies such as the CD-ROM (Compact Disk—Read Only Memory). These technologies are seen by Professors Straub and Wetherbe as being "indispensable for the storage of the massive amounts of information being communicated in the 1990s."

When the Philips and Du Pont Optical Company of Wilmington, Delaware, announced plans in 1989 for the first erasable optical disk for computer data storage, it said the disk would have the capability of storing 4000 times as much information as a floppy disk.

The demand for advanced information technology and storage will promote a different kind of company site—the corporate cluster. It will be more cost-effective for companies to group in shared computerized facilities equipped with advanced digitally switched telecommunications systems than build separate offices.

"We are making our buildings smart by ensuring that they will be adaptable to any future technology that our tenants desired," explained Kenneth W. Hubbarb, executive vice president of Gerald D. Hines Interests, a Houston development company.

Clusters with shared tenant service networks will lower communication expenses. Some experts anticipate that within the next decade ISDN will connect 85 percent of all business telephones nationwide, only serving to reduce costs further.

At the same time information technology is reshaping the traditional office, it is giving impetus to a new category of employee in a new work setting.

Take David C. Barker. Barker used to start his workday by dressing for success, kissing his wife and two young children goodbye, and jumping into his car for the commute to the office. He was lucky if he got home at night in time to tuck his daughters into bed.

In 1986, Barker, who manages eight people for a Philadelphia-based computer communications hardware company, changed his work site. Now he gets up, throws on jeans, a shirt, and some old sneakers, kisses his wife and daughters good morning and heads for work—in his basement.

Welcome to the world of the "open-collar" worker.

While there have always been people working out of their homes—craftsmen, field salesmen, customer representatives, and cottage-industry entrepreneurs—many more occupations can be added to that list because of advances in information technology and the computer, which allows for the give-and-take of business, the exchange of information, and the interaction of people from almost any spot in the world.

Not that many years ago, if someone was a manager, he and his staff had to meet at a common site—an office or factory. Now someone like Barker can get his messages to his staff over the computer, the phone lines, and with facsimile machines.

The country is going on-line. Already at least 35 percent of all households have some type of computer, and there are 1.8 million facsimile machines in the United States. (Eight hundred and sixty thousand were sold in 1988 alone.) Going on-line has swelled the ranks of the open-collar worker. In 1988, according to Link Resources, a New York research firm that tracks the communications technology market, 6 million people worked full-time at home. That represents 5 percent of the American work force. If part-timer workers are included, the number of open-collar workers jumps to 25 million, or 21 percent of the work force.

In 1989 an additional 5 million joined this at-home brigade. "We're clearly beyond the novelty stage," Gill Goradon, a New Jersey consultant who aids corporations in setting up telecommuting programs, has said. "The manager of today's . . . office workers shouldn't have to stand there overseeing people the way the farm boss did with people in the field."

Certain types of jobs lend themselves to telecommuting. Generally these jobs need to have several of the following characteristics. They should entail handling information and have high telephone usage. There should be little need for personal contact. The jobs should be deadline-oriented and be performed mostly at computers.

Some of the occupations more conducive to telecommuting would be:

        word processors
        data entry clerks
        bookkeepers
        secretaries
        computer operators
        computer programmers
        purchasing agents
        bank officers dealing
           with credit and finance
        accountants
        insurance agents
        architects
        engineers

lawyers
writers
travel agents
real estate agents
catalog order clerks
reservation clerks

Telecommuting can be beneficial for both the employer and employee. On the company side, studies have shown improved productivity. Where quantitative measures were possible, productivity went up as much as 100 percent, although 20 percent was more normal. In qualitative studies, open-collar workers said they felt they accomplished more because there were fewer interruptions.

Employers also find that offering telecommuting positions can give them a recruiting advantage. The opportunity of working at home, or at least from a nearby satellite office, is attractive to some would-be employees. A company would have a greater field of candidates from which to choose, rather than limiting itself to people who could commute to the office or were willing to move.

Telecommuting may lead to lower employee turnover. In this age of the two-income family, one of the spouses, faced with having to transfer, might choose to quit instead. Being given the opportunity to work from home might keep someone trained and valuable on the payroll. For example, employees can continue working while on maternity or paternity leave.

On the employee's side, the benefits include having more control over work-home scheduling. Parents, for example, can arrange their day so they have time to be with a child just home from school. Home-work also means savings on food, clothing, and transportation.

Many of those most attracted to open-collar employment are those who want "a way to integrate the pastoral lifestyle they hoped to have in the 1960s with the aspirations to succeed in the business they developed in the 1970s and 1980s," wrote Sherry Turkle, an MIT sociology professor, in her book *The Second Self: Computers and the Human Spirit.*

As telecommunications become more sophisticated, there will be a concomitant increase in telecommuters. As one fan of work-at-home asked, "Why commute to compute?"

But there is more to American business than boardroom and offices, be they downtown, suburban, clustered, or in the basement. This ability to interact through computers will not only change the white collar world, it will change the factory floor, as well. Factory production will have a speed and flexibility heretofore unknown.

At one time, U.S. companies could get away with being monolithic producers—that is, an attitude of "here is the product, take it or leave it." Now they must deal with different markets that often have different needs and demands.

To meet these new demands, American manufacturers have to adapt to more flexible production systems. Traditionally, one assembly or production line was set up for each item in a factory. A different product, or even a modified one, had to be run on a completely separate line. But with the changing rules of commerce, small-batch production must be addressed more efficiently.

Many companies, such as General Motors, are turning to computers to produce the necessary fast modifications. One GM plant, in Somersworth, New Hampshire, manufactures two thousand versions of its basic meter component.

To accomplish such quick-change manufacturing, computer management systems are being integrated with high-speed production on factory floors. Using satellite and telephone link-ups, front offices will have direct control over what goes on in the factory instead of the present time-consuming chain of command, filter-up, filter-down style of management.

Direct link-up systems were not possible before the advent of fiber optics. Factory machines generate high electronic noise levels that scramble computer instructions. Messages sent over fiber optic systems, on the other hand, are immune to the electronic interference.

More complicated and responsive software is being developed that will allow for greater efficiencies. Every nook and cranny of the factory will be monitored by computers. All tools and equipment in the production process will be tracked and controlled. Managers, at the touch of the keyboard, will be able to find out how often a machine has needed maintenance, the movement of all inventory, which tools get used and when.

Monsanto is one company that has already invested heavily in developing process control applications. It expects that the market for integrated process systems that adjust chemical and material flow should be nearly a billion dollars by the late 1990s.

Computers will not solve all the problems of the factory of the future. For example, Corning, Incorporated, had had a lock, through basic patents, on honeycombed ceramic components used in catalytic converters. But with the patents about to expire, Corning was suddenly faced with being undercut by a Japanese company, NKG, that had been making the same components through a licensing agreement.

As Norman E. Garrity, Corning's senior vice president for manufacturing, explained, "In the past we didn't like to get down and dirty in manufacturing. We liked to hide behind patents."

But suddenly NKG was building a U.S. plant, and Japanese automakers were opening factories in America. Corning did not want to lose out to its former licensee, yet it was faced with the problem of each auto company's wanting small quantities of parts based on their own specifications.

To meet the challenge, Corning decided it needed a work force trained in several skills and able to switch assignments as needed. This experiment in flexibility was tried in a Blacksburg, Virginia, plant that had been mothballed. In the revamped factory, there were only two worker classifications—operations associate and maintenance engineer—as opposed to forty-nine classifications in a plant making similar items in Erwin, New York. The experiment started paying off when the company got a contract to supply parts to Honda's Marysville, Ohio, plant.

Of course, it will be easier for other companies to adopt this multiskilled labor force approach if the skills were easier to learn. With that in mind, engineers are turning their attention to something called design for manufacturability (DMF).

Kodak used DMF techniques when designing its Desktop Microfilmer, a low-end office copier. The goal was to have a good product that would be easy to assemble. To achieve that, engineers reduced the number of parts in the machine and made sure that they could be fitted together with few adjustments. For instance, unlike older models where shafts and bearings had to be moved from one side of the interior to the other before it could be fastened, everything in the new model was dropped into the top of the casing onto brackets.

The objective, according to Bob Cole, who was part of the design group, was to "put all the parts in a bag, shake it up and out comes a finished product."

With the new copier, Kodak found it once again had something it could sell on the world market.

Curiously, demographics of the nineties will contribute to the demand for designed for manufacturability products and systems. The American labor force is getting older as the Baby Boom passes. Far fewer cheap new workers will be available. Businesses will have to adapt and use their existing staffs more efficiently. Even McDonald's has seen the future and is buying labor-saving devices such as double-sided grills.

Many American companies have put off new automation expenditures. It was cheaper to hire entry-level workers than pay for expensive robots or new machinery. The days of coasting on high-school graduates are over, however, as there are going to be far fewer of them. Heavy capital will have to be invested in automation.

The term "automation" often conjures up pictures of factory floors devoid of humans, populated only by whirring robots.

Steve Jobs, founder of Apple Computer, has made that picture a reality at his new company, Next, Incorporated. The assembly line at its Fremont, California, factory is inhabited solely by robots that make the "motherboard" of Next's $10,000 computer. Laser "inspectors" check every part, then monitor the accuracy of the assembly.

In explaining why Next went totally automated instead of searching for cheap Asian labor as other computer manufacturers did, Randy R. Heffner, vice president of manufacturing, said, "If you're going to be a global competitor, you have to be able to compete in manufacturing in a world-class way."

Despite Next's enthusiasm for robot assembly lines, robotic applications in most industries is limited. There were glowing predictions for robotics in the fifties, but it has still not lived up to its promise.

According to Raj Reddy, director of Carnegie-Mellon University's Robotics Institute, "Robots are significantly better than thirty years ago, but that doesn't necessarily mean we are anywhere close to [*Star Wars'*] R2D2 or C3PO."

Robots can now make out forms and shapes and measure distances. They can use computers to register thousands of words. They have detectors for smoke and fumes. They can move, usually by rolling. And they recognize textures. In essence, robots can see, hear, smell, move, and touch. The drawback is that so far no single robot can perform more than one or two of these functions.

"Robots are still dumb," Raj Reddy has explained. "They're dumb because we haven't taken the trouble to put the pieces together, not because we don't know how. That takes money and time and effort, and we don't have the money."

One industry highly suitable for even today's dumb robots is automaking. The Toyota luxury car, Lexus, is put together by ninety-four robots in its Tahara, Japan, plant. Lasers inspect the work and report any flaws to humans.

In 1988, when Chrysler switched its Newark, Delaware, plant from the manufacture of the K-car (Plymouth Reliants and Dodge Aries) to the newer Dodge Sprints and Plymouth Acclaims, it spent $205 million on robots, computers, and other machinery. It is interesting to note, however, that Chrysler also opted for a flexible work force, as well. Workers are trained for more than one job, and get more pay according to the number of skills learned.

It has been estimated that 40 percent of the 33,000 robots working in the United States in 1989 were in auto plants. But that statistic is likely to change in upcoming years as demographic pressures put more of an emphasis on automation. Money will be spent on robotics research, existing capabilities will be integrated and new capabilities will be found. Nanotechnologists are even speculating about future miniature "assemblers" building products from the atom up. Nanotechnology, an offshoot of miniaturization, is still a new discipline, but K. Erik Drexler in his book *Engines of Creation* does not rule out that the day will come when nanotechnic factories turn out every product conceivable from food to cars merely by assembling atoms in the proper structure.

It would be no exaggeration to state, in summation, that computers, on-line communications, new flexible job descriptions, work at home, and the upcoming unparalleld level of automation will produce the greatest changes in the worker and the workplace since the dawn of the Industrial Revolution.

While the foregoing has been a look at trends and forces that will affect the country's business climate in years to come, it is possible to look more specifically at certain industries and businesses that will be growing in the nineties and beyond. The following is certainly not an exhaustive list; it does, however, include many of tomorrow's most dynamic industries.

Aquaculture   Toward the end of the eighties, curious stickers began appearing on packages of catfish, stickers that would seem more suited for ground beef or oven-ready roasters. "Farm raised and grain fed" the stickers read. Fish farming—the controlled cultivation and harvest of water plants and animals—is becoming big business as natural stocks of clams, oysters, and certain fish species are depleted.

In 1989 it was announced that five indoor fish farms would be built in Maryland as a joint American–West German operation. According to Michael Gould, the president of the joint venture,

Metz America, this type of computerized fish production system has been used in several European countries, but never before in the United States.

"It is a closed, looped, recirculating system," he explained. "It continuously uses the same water, so it is producing 100 percent natural fish—no chemicals. It's all computerized and very high-tech."

Gould predicts that the five farms will eventually produce millions of tons of fish annually.

In Japan, *nori*—the seaweed used to wrap sushi—is already a $1 billion industry. The potential for seaweed and algae goes beyond a food source. There is talk of converting them into fuel and specialty chemicals.

Auctioning—While the astronomical prices paid for Van Goghs and Picassos at Sotheby's make the headlines, many businesses are finding that auctions are convenient and profitable tools in asset management. Some businesses expected to put more reliance on auctioning include heavy equipment manufacturers, real estate developers, and automobile dealers.

Ceramics Mankind has progressed through the Stone, Bronze, and Iron Ages. It is predicted that we are now entering what could be dubbed the Man-Made Age, with fine ceramics being one of its more important materials.

New high-tech ceramics are only distantly related to tea cups and crockery. This new material is lightweight, can withstand extremely high temperatures, and does not wear out. It is made from artificial substances that are akin to alumina, sand, and titania.

The Japanese government targeted new materials technology in 1980 and put $400 million into a ten-year research program. In a 1982 survey of two thousand Japanese scientists, fourteen of the twenty-two categories cited as being key technologies of the future were dependent on breakthroughs in new materials.

According to Albert R. C. Westwood, director of research and development at the Martin Marietta Corporation, the Japanese "targeted ceramics and are going after it, just like they did with motorbikes, videocassette machines, and other industries."

It is expected that the auto industry will begin foresaking metal for ceramics in coming years. Nissan Motors' 300ZX sports car currently comes with a ceramic turbo rotor. Isuzu and Toyota are predicting their diesel engines will be ceramic sometime in the 1990s. Because ceramics need no cooling or lubrication for temperatures up to 1,500°C, there is an increase in fuel efficiency by 30 to 40 percent.

Ceramics will make an impact on the electronics industry, as well. They are rapidly gaining acceptance in silicon chips.

Child Care A generation ago, only one of every five preschoolers had a working-out-of-the-home mother. By the end of the eighties, the mothers of half of all preschoolers—approximately 13 million children—were away-workers. Parents were spending $14 billion a year for child care and scrambling to get whatever they could as waiting lists grew faster than facilities could get started or expand.

But the story does not end there, it is expected that in the nineties, 60 percent of those entering the labor force will be women, and 80 percent of them will be of childbearing age. Clearly, the demand for child care will continue to rise. The present shortage of good day-care facilities is attracting entrepreneurs who are establishing franchise operations. The shortage will also give a push to companies establishing on-site care centers.

Coatings In vapor disposition or applied coating technology, dissimilar substances such as silicons, metals, polymers, cloth, and even paper are bonded tightly together. The rapid growth of fiber optics, in which this technology is used, will translate into rapid growth for this industry as well.

Computers Many computer companies experienced severe difficulties in the eighties. A drive along the Silicon Corridor, Route 128 near Boston, gave evidence of the distress as companies that had been flourishing were out of business. But this is not to say that the computer industry is dead. Rather, as David Lampe, assistant director of MIT's Industrial Liaison Program, explained in 1989,

"What's happening . . . is the inevitable cycle of what happens in every business."

There are some sectors of the computer industry that will be high growth and dynamic. Small business computers, priced at between $5000 and $25000, are expected to continue to sell extremely well. There will also be growth of linked desktop computers, creating large markets for companies that make equipment for local area networks (LANS) as well as those that manufacture desktops with microprocessors. LANS link computers within limited areas such as a building. For small businesses this is a more cost-effective system than using mini-computers which are terminals linked to a central processor. In 1989 10 million PCs were connected to LANS throughout the world. The International Data Corporation of Framingham, Massachusetts, has predicted that by 1993 that figure will rise to 30 million, or half the world's installed PCs.

Other PC advances expected to make an impact on world markets is what Stephen Jobs calls the "fourth wave." Job's next computer offers a wide spectrum of simulated educational programs. Special features include stereo sound, fully integrated video without distortion, erasable optical disks. The "fifth wave" of PCs will have software systems that talk.

Consulting More companies are turning to outside consultants for advice in strategic planning, human resource management, marketing systems, and use of new technologies. For many businesses, maintaining an in-house staff is not cost effective. Consultants who can get the glitches out of computer systems and make them understandable will be in particularly high demand.

Data Protection It used to be that stealing state or rival company secrets was a question of breaking and entering or bribing a clerk to get at files. No more. Files are kept electronically, and most electronic systems, printers, word processors, and computers can be monitored and then decoded from a distance. This has created the new field of "data protection."

Systems must be tested to see how vulnerable they are to electronic eavesdropping. Once that is ascertained, procedures to prevent the secondary emission of electronic "noises" that can be collected by unauthorized parties are put into place. As companies become aware of their vulnerability, there will be an increased demand for computer security companies.

Drug Delivery Systems   These are ways of more effectively getting drugs into the patient's system. The drugs are not being changed, only how they get into the body. Skin patches are one type of delivery system. They are being used in the treatment of a wide variety of problems from heart disease to seasickness. With research being conducted to develop enhancers to make patches even more effective, revenues in this industry are expected to grow from $500 million in 1986 to $2.5 billion in 2000.

Fiber Optics   The old copper wires of the telephone system are dinosaurs in the information era. They do not have the capability of carrying the volume of information that fiber optic systems do. It is estimated that nearly one billion miles of telephone cables now made of copper wire will be completely replaced by fiber optics.

Health Maintenance Organizations   As health care costs skyrocket, health maintenance organizations, as a more affordable alternative, will rise in popularity. Full health care is provided by HMOs for prepaid fees, as long as the patient uses approved doctors and hospitals. A variation on the standard HMO set-up allows the participating physician to work out of his own office rather than a centralized facility. This arrangement tends to add to the attractiveness of HMOs. In 1987, 24 million Americans were enrolled in such an organization. Conservative estimates indicate that number will rise to 33 million by the turn of the century.

High-Tech Communications   One aftereffect of the break-up of the Bell System was the rush by other, smaller companies to enter the specialty phone and accessory equipment fields. Suddenly there was a deluge of phones shaped like Mickey Mouse, Elvis, and pussycats. Phones with answering machines that had memory, monitors, pauses, redials, and mutes. By the year 2000, these should bring in more than $6 billion in revenue, up from approximately $2 billion in 1984.

Advanced cellular phones will make a big impact on the market of the future. AT&T and Panasonic are expected to be the leaders in the field.

Home Diagnostic Testing This was an industry that barely existed in 1985, bringing in only $100 million. But the business is expanding as different types of testing kits are introduced to the marketplace. In the privacy of your own bathroom you can pinpoint ovulation, monitor blood sugar levels, test for rare venereal diseases, urinary-tract infections, colorectal cancer, and a variety of other diseases and afflictions. The projected revenue by the turn of the century is for more than $1 billion annually.

Home Health Care With advances in technology, acutely ill patients no longer have to be hospitalbound. Infusor pumps, temporarily implanted catheters, and intravenous therapy, for example, can be managed at home. The graying of America will add to the growth of this industry. Given the choice, most people would prefer home to hospitals or nursing homes. Medical equipment manufacturers producing easy-to-operate devices for in-home treatment will experience increasing demand. In 1985 revenues in this field were around $7 billion. That is expected to climb to more than $20 billion annually by the year 2000.

Microfilm Despite the proliferation of computers, businesses are still buried under mountains of paper. Paper documents mean storage costs. The increasingly popular answer to this problem is microfilm. Revenues for companies that transfer business records and communications to microfilm should reach more than $5 billion by the year 2000.

Ocean Mining Four-fifths of the earth is covered by water. Much of the world's resources are hidden thousands of feet below the ocean's floor. Bauxite, iron ore, coal, uranium, tin, silver, platinum, and gold are being mined to some extent now. Fullscale operations to recover valuable resources from seabeds should be launched in the next decade. In 1986, approximately $1 billion was spent on equipment for such operations. By 2000, that should hit more than $7 billion.

Programmable Communications Systems   These have the largest potential for growth within the communications field. In particular, manufacturers of advanced switching systems will experience tremendous growth as videotex, electronic mail transmission, and other expanded data base use become more popular. There are some predictions that this field will accelerate from $20 billion in annual sales to $35 billion within the next decade.

Robotics   Though many prognosticators are backing away from saying factories run entirely by robots are what is on the horizon, most still agree that a new generation of intelligent robots will be used for cleaning hazardous waste dumps, for picking perishable fruit, in sophisticated security systems, and for food handling. As sensor technology improves, robotics will have increased applications. The approximate revenue in 1987 was $400 million and will undoubtedly top $1 billion by the year 2000.

Supercards   Supercards are credit cards taken several miles further. Credit cards serve one function. Supercards, with memory capacity of some 64,000 characters along with a digital display, will be multifunctional. For instance, "lifecards" will hold up to 800 pages of medical records and may provide doctors with such information as EKG readings, X-rays, and insurance files. Welfare payments may eventually be made through supercards. Applications will proliferate as the cards gain wider acceptance. It is projected that 2 billion supercards will be in circulation by the year 2000.

Superplastics   In recent years, plastics have been much maligned as being harmful to the environment, but their popularity has not declined. American industry consumes more plastic than steel, aluminum, and copper combined. Plastics are being used in caskets, in electronic components, and in car manufacturing where instrument panels and bumpers are being made out of specialty polymers. In 1980 approximately 150 pounds of plastic were used in a car. By the mid-1990s that should rise to 225 pounds. And, apparently, this is not at the expense of safety. Pontiac used a predominately plastic shell for its Fiero model. The crash performance of the Fiero was rated superior to any other vehicle ever tested in the 2,800 pound category.

Plastics, being artificial, can be custom-designed by scientists for various uses. They have the advantage of being cheaper, lighter, and stronger than metals. This makes them especially appealing to the aerospace industry. Already plastics have been used in the airframe and wings of AV-8B Harrier jump-jets.

Projected revenues in superplastics for the year 2000 are $1.5 billion, up from $125 million in 1986. Du Pont alone expects its specialty polymers business to increase 10 to 12 percent a year, with sales of high-performance plastics growing at two times the rate of the world economy.

Just as there are certain industries that can be tagged as growth sectors, certain occupations will be in high demand in the nineties. The trends and changes brought on by technology, globalization, and demographics will have a major impact on the face of the future job market. People seeking to succeed will find a greater emphasis placed on education, training, and the ability to assimilate new information. Some of the growth jobs and careers will be directly linked with computers and new technologies. Others will be specialized niches within traditional job sectors.

In high-tech technologies, great shortages are expected in applied optics, laser applications, electromagnetics, avionics, and especially advanced composites required for tomorrow's lighter aircraft. Employers will be looking for specialists in polymer science.

Engineers of all types will be in short supply throughout the next decade due both to a 25 percent drop in engineering school enrollment during the eighties and an increasing number of job openings. It is predicted, for example, that in the nineties there will be a demand for 50 percent more electrical engineers.

Companies going global will be hiring additional manufacturing engineers to design factories that meet the demands of the future. These engineers will be called upon to speed up the concept-to-finished-product process by restructuring the factory, designing and operating robots and other machinery, and fully incorporating advances in computer technology.

Law is one traditional niche that will experience a high-demand for specialization. Companies going global will need international lawyers, conversant in the foreign language and culture, as deals are made and ventures structured. In 1989, American companies such as AT&T, Du Pont, and Pepsico poured $7 billion worth of investments into Europe alone.

Companies will also be looking for accounting and financial professionals with international backgrounds.

A demographic challenge employers in the next decades will have to face is the graying of America. As the pool of entry-level employees contracts, firms will have to make adjustments. For one thing, there will be a greater emphasis on retraining current employees. Corporate trainers will be hired on staff or brought in as consultants to do such things as explain computer systems, demonstrate new machinery, or give short courses on foreign countries. They will also be teaching basic skills to entry level prospects who might otherwise be unemployable.

Another side effect of the graying of America and the decrease in the number of younger workers will be the pressure put on companies to hold on to employees. More companies, therefore, will be looking for business-resources managers. These are new-age personnel administrators, charged with keeping employees happy and productive. A business-resources manager might mediate disputes, prod the company into setting up day-care facilities or job-sharing programs, or counsel managers in other departments on how to get the most from their workers.

When we go beyond the nineties into the next century, new job opportunities will open up where there will be a demand for specialized scientists, engineers, architects, technicians, and ancillary personnel. However, we might not have enough people to fill the positions because of a shortage of teachers and professors. An Andrew Carnegie Foundation study predicted that between 1997 and 2002 there would be only seven candidates for every ten professional positions in the humanities and sciences. The ratios in the natural sciences were only slightly higher.

This teacher gap has the potential for disaster. The country will not be able to take advantage of the future economic potential if there is not skilled, trained personnel available. We will be forced to tag along behind other countries' technology if we do not have the scientists and engineers to develop our own. Then the offices and factories of the future will not be in the United States, and instead of Business Not as Usual, it will be Business Not at All.

FOUR

Tomorrow's Magic Carpets

PICTURE THIS: Soaring high above the earth, on the fringe of the atmosphere, the "Orient Express" barrels across the Pacific Ocean at seventeen thousand miles an hour on its two-hour trip from Washington, D.C., to Tokyo. This vision of a "hypersonic plane" traveling at twenty-five times the speed of sound, or Mach 25, is perhaps the most dramatic version of travel now attracting both thought and capital.

Back in 1986, President Reagan featured this vision in his State of the Union speech, and NASA has diligently been researching the "National Aerospace Plane" or NASP. Indeed, more than $450 million in contracts have been awarded for the initial studies on the Mach 25 aircraft. Entire development costs for an actual plane are expected to range somewhere between $8 billion and $24 billion.

The NASP may be the greatest proposal for future travel, but it is far from the only one that will change how we get from place to place. Air, rail, and auto travel will be markedly transformed in the coming decades.

Changes in transportation seem to come not in gradual stages but in leaps. The twenty years between 1945 and 1965, for example, saw much greater changes, with the passenger jet being perfected

57

and high-speed highways reaching out across American and Europe, than either the twenty years before or after.

Now, we are again poised for another leap.

The hypersonic plane (HSP) is out there not only on the edge of the atmosphere, but also on the outer edge of the impending transportation leap. Before such aircraft start carrying commercial passengers, HSPs will have to overcome a welter of technical problems unparalleled in aeronautical design.

"There are horrendous technological challenges," conceded James Arnold, chairman of the 125-person NASP working group at NASA's Ames Research Center. "It's a high-risk program, there's no question about it." Nevertheless, Arnold has said, "In my judgment it's feasible. If it wasn't, I don't think the United States would do it."

The armed forces are also intrigued with the HSP idea, and the Defense Department is helping to finance the initial studies.

Still, the technological hurdles appear to be immense. Such a plane could use neither current jet engines, which are not powerful enough, nor rocket engines, which consume too much fuel. The air friction resulting from flying at such speeds would heat up the plane to temperatures of 11,000 degrees Farenheit. Finally, controlling a vehicle hurtling across the sky at such tremendous speeds would be far more difficult than flying anything ever built.

To date, only the space shuttle has traveled at speeds of Mach 25, on reentry into the Earth's atmosphere. The fastest plane launched from the ground is the X-15A, which reached 4,534 miles per hour, during a 1967 flight. It is hoped that a prototype of the hypersonic plane can be built in the next decade. It has already been dubbed the X-30.

Still, the obstacles before the X-30 are daunting. But as James Arnold has said, "This is one of the most fascinating, challenging problems. And that's what turns technology people on. How do you do something like this?"

There are some tentative answers to Arnold's question. Researchers believe that the hypersonic plane will need two sets of engines, one to boost it to Mach 5 and a second to carry it to higher speeds. This second engine is, however, still in the experimental stage.

Problems of heat could be dealt with by using heat-resistant silicon-carbide composite materials in building the aircraft, a hydrogen cooling system through the plane's engines and body, and unique revolving mechanisms in the nose and along the wings to absorb and distribute heat. The nose of the plane, for example, would act like the ball in a roll-on deodorant, spinning as the plane moved forward. Cylindrical rolls on the front edges of the wings would also spin as the plane flew. These revolving mechanisms would help dissipate the build-up of heat.

As for piloting the X-30, computers and video screens would have to play a principal role. Computers with artificial intelligence systems would constantly have to monitor the plane's performance and make adjustments. Visual information would have to come from television monitors, because it would be too difficult to install windows in the cockpit. That would enable the flight to be monitored and controlled as easily from the ground as from the cockpit itself.

Nevertheless, it is clear that the technical barriers that the NASP must clear are immense, and the project is not without its share of critics. Richard Shevell, a professor at Stanford University in aeronautics and astronautics, calls the vision of the Orient Express "a fallicious dream," and Stephen Korthals-Altes, of the Massachusetts Institute of Technology, maintains the plane "is being oversold."

What critics and supporters can agree on at the moment is that development of the NASP is definitely not around the corner. This technology is not likely to be sufficiently developed until into the next century.

"The whole idea of the aerospace plane is extremely optimistic and unlikely to happen the way they're talking about in less than twenty-five or thirty years," Sevell has said. NASA has estimated that the plane is still probably forty years in the future.

The Mach 25 plane is not the only new supersonic transport on the drawing board. Work on a more modest, slower plane is also underway.

Both the Boeing Company and the McDonnell Douglas Corp., the only U.S. manufacturers of large commercial jets, have already prepared studies for NASA's Langley Research Center, in Virginia, on the prospects for a supersonic plane, and NASA has concluded that there is a $200 billion market for a new generation of large, long-range, high-speed passenger planes, primarily because of the growth in Pacific air traffic.

"Growth in trans-Pacific travel would justify designing and building a new supersonic passenger airliner and placing it into service sometime between the years 2000 and 2010," according to NASA.

The Boeing study projected that by the year 2000, a fleet of 1,200 subsonic carriers would be needed to carry the estimated worldwide long-range traffic of 315,000 passengers a day. Because supersonic planes are faster and therefore can make more trips per day, it would only take a fleet of 800 such craft with a capacity of 247 passengers to manage the same traffic demand.

These planes, according to the initial designs and proposals, would be more modest than the X-30. The plane would travel at speeds up to Mach 3, more than 2,000 miles an hour. That would still cut the current nine-hour Los Angeles to Tokyo trip from nine to four hours. While the planes would fly in the stratosphere, at altitudes of 55,000 to 60,000 feet, they would not encounter any of the subspace problems in flying an X-30.

To be economical, Mach 3s will also have to carry 250 to 300 passengers and be able to fly up to 7,000 miles without refueling. For it to be competitive, the cost of flying such a plane should be no more than 10 to 15 percent higher than the cost of subsonic planes.

The Mach 3 passenger plane has its own share of technical problems. To have the required cabin capacity and achieve the required speed, a radical new design will be needed.

So far, the most promising version is the "double delta wing." These are broad triangular wings that flair out considerably at their tips. Unfortunately, this configuration will make it harder to keep the plane stable in flight. Engineers are hoping to overcome the stability problem with a new electronic control system.

Like the X-30, Mach 3 planes will need lightweight materials that are heat-resistant. Again, researchers are looking to composites to solve these problems, but the exact materials have not yet been identified.

Because the Mach 3 jets are so much closer to reality than the X-30, engineers are already dealing with political and environmental concerns. For example, aircraft designers must find a way to limit the sonic booms created by supersonic planes, so that they may operate over land. The Concorde's effectiveness on its Paris/New York run was limited by the requirement that it reduce its speed to subsonic levels well before approaching the New York airport. Similarly, the new plane's engines must be quieter than the Concorde's to win local approval for frequent takeoffs and landings.

The largest issue that the builders of such planes will have to resolve, however, is the environmental danger such aircraft might pose to the Earth's ozone layer.

Residue of the burned off kerosene-based jet fuel includes nitrogen oxides, which are believed to play a key role in the destruction of the planet's protective layer of naturally occurring stratospheric ozone. This layer shields the Earth's surface from dangerous ultra-violent radiation from the Sun.

The answer to the noise and ozone concern may be found in the development of a new type of engine, one that burns cleaner and that operates efficiently at both sonic and subsonic speeds.

Unlike the efforts to develop the X-30, the research to find the necessary components to build these Mach 3 planes remains totally in the private sector, and it may prove too large a job for any one company.

"In the past, building a new airplane meant you had to bet the company," explained James P. Loomis, director of the Center for High Speed Flight at the Battelle Memorial Institute, a research organization in Columbus, Ohio. "This effort will in all likelihood mean that four or five companies will have to join forces." Negotiations are currently underway to form a joint venture to develop the

Mach 3.

NASA is anticipating that the emerging market for such a plane will spur on American industry. "This market," NASA announced, "could be an important factor in the U.S. sustaining a positive trade balance in the aerospace market." But the agency added that it will be "up to industry to build and sell high-speed, civil transport on their own."

The market for the Mach 3 appears to exist, so the question becomes whether the technical problems, which remain much more modest than those of the X-30, can be solved. In all likelihood they will be, making the prospects bright for a new generation of supersonic passenger planes.

American companies may find they have competition in the Mach 3 development. Europeans, who already have experience with international consortiums for building commercial aircraft, such as the Concorde and Airbus, may eventually begin a similar project. There is even the possibility that the Japanese, who also have great experience in coordinating industrial efforts through joint ventures, might get involved in order to enter the aircraft industry.

One field of air transport where the United States enjoys a lead on the rest of the world is in the development of the tilt-rotor aircraft, which is part helicopter and part airplane. And like the Mach 3 jet, it appears to be a technology with a waiting market.

The aircraft, built by Bell and Boeing, was originally designed for the Marine Corps and is known as the V-22 Osprey. The Osprey is powered by huge rotor blades on each wing tip. During takeoff the blades are in a horizontal position, enabling the aircraft to lift off like a helicopter. Once in the air, the blades swivel into a vertical position so it can fly like a turboprop plane.

The Marines had planned to use the Osprey to replace helicopters, but at a finished cost of $22 million each, the military deemed the Osprey too expensive. For nearly two years the Bush administration has tried to scrap the Osprey. But Congress has continued to fund the program, allocating as much as $225 million in 1989 to continue research and development.

While the vehicle may not be what the Marines need, it appears that there are a lot of intercity commuters ready to climb aboard. Since the V-22 is able to land in densely populated areas, like a helicopter, but cruise like an airplane, it is ideal for jumping, say, from Wall Street to Capitol Hill in an estimated forty-five minutes. The advantage is that there is no need to fight traffic to and from airports.

A NASA study has predicted that by the year 2000 there could be as many as 1,600 tilt-rotor aircraft operating between U.S. cities.

Tilt-rotors can carry up to forty passengers from downtown to downtown faster (they travel at speeds of up to 350 miles an hour) and more conveniently than either helicopters or airplanes for trips of 300 miles or less.

Proponents of the tilt-rotor foresee a network of "vertiports" located in or near the center of major cities. Besides providing transportation for short-hop travelers, the introduction of the tilt-rotor and its vertiports will ease congestion at metropolitan airports, which are now choked with the short-hop traffic. For example, at the New York City area airports—Kennedy, Laguardia, and Newark—40 percent of the flights are to destinations of 300 miles or less.

The Federal Aviation Administration has already awarded $2.7 million in grants to fifteen cities and states to study the feasibility of building vertiports. "Nothing on the horizon offers the potential for capacity increases system-wide like the tilt-rotor," said Richard Milhaven, a planner with the New York–New Jersey Port Authority.

Development of the tilt-rotor is not without its drawbacks. First, it is a relatively expensive aircraft, with estimates for the civilian vehicle ranging to $18 million, as compared to $9 million for a comparable turboprop. Similarly, maintenance on the tilt-rotor, at about $835 an hour, is almost five times as expensive as a regular turboprop.

Figures of this magnitude have made the airlines somewhat wary of the craft. "The number of passengers paying for it may not justify the service," said Steve Horner, a regional airline analyst with

Avmark, an aviation consulting firm based in Arlington, Virginia. Still, the advocates contend that the costs in time and travel to and from airports will lure intercity travelers.

"You also eliminate the aggravation factor by not having to go through an airport. There are a lot of people who would be willing to pay for that," said David Jensen, editor of *Rotor and Wing International*.

One of the strongest indications of the future of the tilt-rotor is the interest shown by those who do not yet have the technology. Eurofar, a consortium of companies from Great Britain, France, Spain, Italy, and West Germany, has been formed to develop a similar vehicle. A Japanese company named Ishida reportedly hired several former Bell engineers to work on their own project.

"Right now, we have a lock on the technology. We don't want to wind up like the VCR," warned Mark F. Fling, a Los Angeles-based aviation financing consultant, in reference to the American-developed but Japanese-exploited videocassette recorder.

The tilt-rotor is not the only answer to the growing congestion of intercity travel. Trains will also play a greater role in the future. But in this area the technology will be either European or Japanese.

Both the French and the West Germans have been leaders in the push to develop high-speed trains. The French have opted for using existing technology and improving speed by enhanced design and planning. The French TGV (Le train de très grande vitesse) travels at speeds of more than 186 miles per hour, relying on electric engines and sophisticated roadbeds built for speed.

Already the TGV is speeding between Paris and Lyon and Bordeaux. There are plans to send the train all over Europe and even to Great Britain through the "chunnel," the tunnel being built under the English Channel.

Meanwhile, the West Germans have been pushing the edge on both speed and technology with their "magnetic levitation" train, or simply mag-lev, which uses magnetic energy to hover over a rail guideway and can travel at speeds of up to 300 miles an hour.

One of the biggest brakes on the speed train movement in the

United States has been the cost. Estimates have run from $500 million for just a twenty-mile route to $10 billion for 400 miles of track. In Europe, much of this is financed through government programs, but in the U.S. it most likely will be paid for privately.

Nevertheless, with auto traffic becoming more and more inpenetrable and airport congestion growing, high speed trains do offer an alternative. In France, to offer one example, one can get on the Paris Metro, go to the Gare de Lyon, jump on the TGV, get off in Lyon, take the Lyon subway, spend a full business day, repeat the exercise in reverse and be home for dinner.

And so, based on the appeal and the need for new transportation alternatives, both the French and the Germans are in pursuit of the American market. The Morrison Knudsen Corp., representing Alsthom TGV, the French manufacturer of the TGV, has proposed building a 245-mile, high-speed train line between Houston and Dallas. It would cost about $2.2 billion and link Texas's two largest cities by a ninety-minute train ride.

The TGV Co., another Althsom-backed firm, is seeking to build routes in Florida to connect Tampa, Orlando, and Miami.

The German Federal Railways have also considered proposing a high-speed line, similar to the TGV, for the Dallas–Houston market. This proposal mag-lev project is called the Intercity Express, or ICE.

It is with the mag-lev that the West German's are hoping to capture key elements of the American rail market. Transrapid International, the West Germany company building the mag-lev, has been eyeing potential routes in Southern California and Florida.

California and Nevada's legislatures have created a bistate commission to look into the feasibility of a Los Angeles to Las Vegas high-speed train.

Florida, with its burgeoning population and booming tourist industry, however, is the first big battlefield for high-speed lines. By the late 1980s, the state was attracting 50 million tourists a year, and that figure is expected to expand during the next decade to 90 million.

High-speed trains are seen as one way of dealing with traffic and airport congestion, and that is why the Florida high-speed rail commission has been the prime target for foreign railroad interests.

In addition to the TGV and Transrapid proposals, the Florida commission has also received proposals from the Florida High Speed Rail Corp., which represents Asea-Brown Boveri, a Swiss-Swedish company that is one of the world's largest builders of rail cars, and from Mag-Lev Florida, backed by American and Japanese investors.

The Florida commission is interested also in construction of a twenty-mile route that would link the Orlando Airport to Disney World. Mag-Lev Florida has already acquired the rights on the land it would use to build its elevated rail system.

While Mag-Lev Florida's backing is Japanese, it has indicated that if it gets the state's approval for its plan, it will go with the Transrapid International technology, which appears to be at least several years ahead of the Japanese mag-lev.

Indeed, there is already a 19.5 mile prototype for the German mag-lev operating in the Emsland farming region of northwest Germany, and in 1989, the West Germans set out to build a ninety-mile line between Hanover and Hamburg, which is expected to be operating by the mid-1990s.

With its airplane-styled cars, quiet, smooth ride at speeds of up to 300 miles per hour, Transrapid executives believe that the mag-lev will eventually make all other rapid train systems obsolete.

Despite the emergence of supersonic jets, helicopter-airplanes, and speed trains, most Americans will still probably insist on getting around on their own, in their own vehicles.

While the car of the future will have a more efficient engine, use different fuels, and have better pollution-control equipment, perhaps the most striking change will be in its ability to help you get where you want to go. Cars of the future will have navigation systems and electronic maps.

Such systems and maps are already widely used by ocean vessels, and today approximately one thousand Southern California motorists have a $1,495 electronic map called Navigator in their cars.

General Motors is considering making an improved version to be called Travel Pilot, which will be offered as optional equipment in its 1993 Oldsmobile.

Making computer maps extremely appealing is the fact that the U.S. Census Bureau has recently compiled the most detailed maps of the country every made, and they are all on a computer system called TIGER.

Coupled with nagivation computers that determine your exact location from satellite signals, the maps can help a driver know just where he is and where he is heading at all times. With just the push of a button, the driver can see where he is on a statewide map, a city map, or a local street grid. When computer interfaces get more advanced, the device will verbally warn the driver when he should take a right or that there is traffic ahead and an alternate route is advisable.

Computer systems can also be used for sending distress signals. Some trucking companies are already using computer messaging. With the push of a few buttons, a truck needing repairs can alert its dispatcher where it is and what it needs, even if the broken-down truck is 1,500 miles away.

General Motors, the Federal Highway Administration and the California Department of Transportation are conducting a $1.6 million experiment called Pathfinder that attempts to provide two-way communication between motorists and traffic information centers. Current road conditions will be displayed on the computer maps and alternative routes suggested.

Pathfinder is seen by those involved as more than just a new convenience for the American motorist. "Traffic congestion is a major social and economic issue in most metropolitan areas—and it's going to get worse," said Walter Albers, GM's director of the Pathfinder project.

Unless something is done, warned Frank Mammano, the Federal Highway official overseeing Pathfinder, we face "a bleak future of rapidly worsening traffic congestion and a serious loss of mobility and negative impacts on the nation's economic vitality and quality of life."

FIVE

Old MacDonald's New Age Farm

THERE IS NO understanding the future of agriculture without first going back to the beginning.

We must look back to those Neolithic women who tired of waiting for their men to return to the cave with some meat, tired of foraging for nuts and berries and moving on when the branches were bare, tired of the feast-or-famine existence that was life. Rather than sit and moan about their fate, they did something. They became the world's first agricultural genetic engineers.

As Norman Borlaug, Nobel Prize winner for his work with hybrid wheats, once explained, these early female pioneers were "responsible for all the bread and cereal products that we eat today. Neolithic woman was the greatest plant biologist in history. She understood that she had to domesticate wild plants and improve them."

Basically, these women did what farmers and scientists have done through the centuries—they selected. They chose the plants with seeds that grew well and quickly, that were easier to harvest, that tasted best. The seeds from the desirable plants were used for the next generation of growth, and so on.

Taming the plant world, or at least part of it, went hand in hand with animal domestication. As century followed century, farming techniques became more sophisticated.

By the mid-twentieth century, modern agriculture probably seemed far removed from those prehistoric women scratching in the soil. The years following World War II had ushered in the Chemical Age of Agriculture in which more and more reliance was put on pesticides. In 1948, 15 million pounds of pesticides were used, by 1989 that figure had risen to 125 million.

Some farmers, such as Texas cotton growers, became almost totally reliant on pesticides. According to Dr. Perry Adkisson, vice-president for Agriculture and Renewable Resources at Texas A&M University, "It was like magic. Farmers planted longer-fruiting cotton and made unheard-of yields under an umbrella of insecticides."

Chemical pesticides and fertilizers had opened the door to the Green Revolution. Biologists were able to concentrate on increasing the plant's yield. Crossbreeding and hybridization became state of the art farming. It was touted that so great would be the effects of the Green Revolution that the world would soon be free of hunger. India would go from a food importer to a nation that could feed its vast population with its own hybrid wheat and rice. Mexico, Pakistan, Turkey, and the Philippines would all emerge from under the shroud of famine and misery. (That it didn't quite work out that way is one more tale of high hopes and misguided technology.)

Yet for all the high technology and advanced science involved with high-yield hybrids, chemical pesticides and fertilizers, modern agriculture was still a variation on the Neolithic woman's theme. As Jack Doyle put it in his book *Altered Harvest—Agriculture, Genetics, and the Fate of the World's Food Supply,* we were doing what always had been done, "standing outside the process, watching and reacting to the natural growth of plants and animals. Even during the industrial and scientific revolutions of the last one hundred years or so, man has improved his crop and animal lines by carefully observing and recording how and what they produced, and then breeding the best of them to obtain selected characteristics."

For better or for worse, we are now changing venue. No longer will we be standing outside, we have crossed the threshold and have gone inside the plant.

Thus begins the Era of Biotechnology, of genetic engineering.

For the first time, man is capable of going inside the plant to alter its genome. By manipulating their genetic makeup, we will be able to produce plants that grow faster and bigger, have built-in, natural pesticides, don't need as much water, and fertilize themselves.

Biotechnology has thrust beyond theory and the laboratory into practicality and the test field. The next step will be to the super-market shelf.

Before making biotechnology a reality, scientists had to find a way to introduce new genetic material directly into plants. And as is often the case in science, the most improbable hero—much like that moldy bread and penicillin—came riding to the rescue. In this case, the hero was a bacterium, one which causes tumorous growths, called crown galls, to appear on plants. When the bacterium makes its way into the plant, its genes become part of the host's genes. All the genetic engineers had to do then was splice the gall-producing gene out of the bacterium and replace it with a more desirable one.

In one early test, a gene from a pea plant was successfully put into the leaf cell of a petunia. (Petunias and pea plants are con-sidered laboratory mice of the plant world.)

Other techniques for introducing DNA into a plant's protoplast are being developed and refined. One technique uses a particle gun that virtually shoots that gene where you want it to go. Using pollen to transfer genes is also being studied, as well as the rehydration of dried-up embryos and injecting the new gene directly into the plant's reproductive organ or into immature embryos.

It is predicted that new genetically altered corn, soybean, rice, cotton, rape oilseed (used in cooking oil), sugar beets, tomatoes, and alfalfa will hit the market sometime between 1993 and 2000.

Scientists are most concerned with developing plants that are resistant to insect and disease, and with creating new ways of deal-ing with weeds. In the past, we sprayed crops with insecticides, saturated the soil with weed killers, and crossbred plants hoping for hardy offsprings that could withstand disease. Being able to go into the crop plant's genome offers much more flexibility to tailor-make the organism.

In the area of weed control, for instance, a multimillion dollar race is on to start selling seeds for herbicide-resistant plants. A fine line must be walked when applying herbicides. Too much will destroy the crop along with the weed. These new plants would be able to take a bigger dose of the herbicide without being damaged. It is probably no surprise that some of the big herbicide companies are funding this research. For instance, American Cyanamid backed the development of a corn plant resistant to a new American Cyanamid weed killer. The company then sublicensed the rights to the plant to Pioneer Hi-Bred International, a Des Moines, Iowa, seed company. Pioneer hopes to get the seed into farmers' fields early in the 1990s. The farmers then could use the Cyanamid chemicals to kill undesirable grasses that choke corn, without harming the corn.

There are those who strongly oppose genetically engineered herbicide-resistant plants. They argue that at a time when herbicides are in the water table of most of our farming states, why grow plants that would allow more herbicide use? But proponents counter that these will be new herbicides, more effective and safer for both the environment and animals. For instance, Monsanto projects include producing soybeans, tomatoes, cotton, and corn that would be glyphosate-resistant. Glyphosate is a Monsanto herbicide that the Environmental Protection Agency calls one of the safest agricultural chemicals used today. It does not pollute the water and does not stay in the soil long.

Nonetheless, people like Jack Doyle, who is also director of the Washington-based Agriculture and Biotechnology Project for the Environmental Policy Institute, are concerned. He has said that using biotechnology techniques to produce herbicide-resistant plants is "troubling because they suggest a continuation of chemical toxicity in agriculture and the extension of the pesticide era."

Although scientists say that someday plants will be genetically engineered so no chemicals will be needed at all, in the meantime, as Tufts University professor Sheldon Krimsky maintains, "The ideal thing for some chemical companies is to make everything dependent on chemicals so that when you buy the seed you have to buy the chemical."

Less controversial is the development of disease-resistant plants. In the past, hardy plants were developed by crossbreeding with wild varieties having natural resistance. This took years and was often unsuccessful. For instance, years were spent crossing a wild South-American tomato resistant to mosaic virus with an American commercial variety. The result was a fruit that neither viruses nor humans liked. With genetic engineering, it will not take years to find out the tomato you are producing has an unappetizing taste. If a genetically engineered plant does not live up to expectations, scientists can quickly move on to the next test. Also, with gene technology, a new trait is being introduced to the plant while the old traits remain. That lovely wait-for-it-all-year taste of a Jersey tomato will still be there once a disease resistance is introduced.

The important crops expected to benefit from this type of gene manipulation are corn, wheat, vegetables, rice, and soybean.

With insect-resistance, genes for proteins that are actually lethal to certain types of insects will be put into the plants. So instead of having to spray chemicals, the plant itself will ward off or kill off the insects. Of course, the trick is to find proteins that only kill the bad bugs and that have no ill effects on good bugs, other animals, and humans. This may sound as if the impossible is being asked, but one such protein, the Bacillus thuringienis (B.t.), has been identified. This protein is a poison pill for moth and butterfly larvae.

It is expected that leafy vegetables, corn, and cotton genetically engineered to be toxic to caterpillars will be the first on the market.

Of course, genetic engineering will not be restricted to disease resistance and weed control. The potential benefits of manipulating the plant's genome are staggering. Canners will be able to package tomatoes that are more solid, with less water to extract. Dwarf fruit trees will be developed to make harvesting easier.

There has been talk of crunchier carrots, crispier celery, and pop-corn that tastes so good no one will want butter. There are certain things, naturally, that purists would prefer left alone. Beer, for example. Something called the Brewing Research Foundation of Surrey, England, is working on genetically engineered beer. The foundation is trying to reduce starches with an altered yeast. One taster of the gene-spliced brew volunteered that "It tasted precisely the same" as any other light beer, which, to many beer drinkers, may not be much of an endorsement.

Cheesemakers are facing a shortage of rennin, a substance that comes from the stomachs of calves and is needed to curdle milk. Dow Chemical bought the rights to a technique that grafts the genes for calf rennin into yeasts and bacteria, promising an uninterrupted supply. Purists have again raised objections. They believe that some of the good taste in cheese comes from impurities in calves' rennin. To compensate, scientists may genetically engineer the impurities, as well.

There is an inevitability about this great change in agriculture. It will happen, but the question is when. Two factors affecting the speed with which genetically altered crops will end up in the produce department and on store shelves are the regulatory process and public acceptance of this biotech wizardry.

By September of 1989, only thirty field tests had been approved by the United States Department of Agriculture. It took years and a daunting application process to get that approval. Some researchers threw up their notebooks in frustration and backed away altogether. As John Sanford, a Cornell University researcher, once explained, "People hold off when they know they're going to have to submit a document as thick as the telephone book to get approval."

It took Steven Lindow, of the University of California, Berkeley, three years before he was able to spray his genetically engineered frost-fighting bacteria on a field of potatoes in Tulelake, California. He had to negotiate a maze of regulation that involved the Department of Agriculture, the Food and Drug Administration, and the

Environmental Protection Agency, all of which are still groping for workable guidelines.

There can be no question that strong regulation is necessary. One does not have to look further than the "but that will never happen" mentality of the nuclear energy proponents. For years they promoted nuclear power plants as being clean, efficient, and safe. Then came Three Mile Island and Chernobyl, and somehow the claims did not carry much weight anymore. That was an industry reputed to be highly regulated.

There are legitimate fears associated with modifying Mother Nature. Washington activist Jeremy Rifkin has voiced concerns over potential disasters. He filed a suit trying to stop the testing in Tulelake, arguing that the bacteria might alter the ecological balance and change the climate.

Wes Jackson of the Land Institute in Salina, Kansas, is another voice urging caution. "Ninety-nine of these experiments will be harmless, and then one will be in the category of ozone destruction."

As mentioned previously, other environmentalists have expressed specific misgivings over increased use of chemicals with herbicide-resistant plants. Well-defined regulations must be formulated to insure testing of environmental impact, to guard against the new plant's presenting a direct risk to human and animal health. But at the same time, a reasonable regulation process must be instituted so that companies have a fair idea how long it will take to commercialize a new plant. Charles S. Gasser and Robert T. Fraley, from Monsanto, wrote in *Science* magazine that "regulation of transgenic plants must be based on scientific principles that (i) meet the general public's need for a safe and reasonably priced food supply and (ii) recognize the inherent low risk of gene transfer technology and the benefits afforded by genetically engineered crops to growers, food processors, and consumers."

It must be encouraging to the industry that the public is becoming less fearful of biotechnology. No one is expecting that, as in some low-budget B-movie, giant cucumbers will rise up to walk the earth and demolish drive-ins. Nonetheless, those thirty field tests have been on a small scale. Regulation and follow-up will be needed once the genetically engineered product goes to widescale use. It won't be known until it's out in the field if some genetically engineered microbe unexpectedly kills an insect beneficial to a certain crop.

What does have the public upset is BST—Bovine Somatotropin, a genetically engineered growth hormone for cows. BST helps cows convert feed to milk faster, thereby increasing milk production by as much as fifteen percent. But the public has seen too many Mike Schmidts praising milk as Nature's best food—you just don't fool around with milk. Some supermarket chains have announced that they will not carry BST milk. It remains to be seen what the reaction to a hog-growth hormone, promising leaner pork with lower feed costs, will be.

For genetically engineered agriculture to take off and fulfill its potential, genetic maps of important crops are needed.

Genes and combinations of genes determine the plant—how much starch it will contain, how big it will be, how fast it will grow, how much water and sunlight it will need, how sweet or tart it will be, how vulnerable to insects and disease it will be. To adjust, rearrange, and manipulate the genome, you have to know where the individual genes are located and how different genes work together.

This is a formidable undertaking. In fact, it is so vast and complex that no single company, no matter how eager to be in the vanguard of genetically engineered agriculture, could afford to handle the whole project. The Japanese government is planning to spend $200 million a year, just on rice genetics. Some European countries are backing projects to map genes of grains and vegetables. In the United States, until the beginning of the Bush administration, it was left to industries and universities to work on partial maps of certain plants. Then in February, 1989, the new secretary of agriculture, Clayton Yeutter, announced that the USDA would begin working on a plan for developing genetic maps of important crops.

His announcement did not mention how much this might be

expected to cost, although it is estimated the human gene map would carry a price tag anywhere from $1 billion to $3 billion.

It was actually the outgoing assistant secretary for science and education, Orville Bentley, who impressed upon Yeutter the need for crop maps. According to *Science* magazine, Bentley showed Yeutter a USDA conference report that concluded "that in order to maintain U.S. competitiveness in world agricultural markets, the government must accelerate the mapping of genes and sequencing of DNA for crop plants."

The report went on to say that partial maps already developed at universities and other research facilities need to be coordinated with federal projects and that the results of all research should be placed in a centralized data bank.

Genetically engineered agriculture may be what makes the future covers of *Business Week* and becomes the lead stories in *Time,* but there will be other developments in food production. Some of them might not be as awe-inspiring and others might seem too far-fetched for reality, but they are waiting over the horizon.

Artificial seeds, for example: one may wonder why anyone would need artificial seeds. What is wrong with natural ones? But consider seedless watermelons and seedless grapes. Now it is a slow process growing these plants from cuttings, but scientists at Florida's Institute of Food and Agricultural Sciences are working on a speedier technique. They are creating "somatic embryos" by culturing pieces of the plant with special hormones and chemicals. These embryos, about the size of regular seeds, are then dehydrated. According to Institute biologist Dennis J. Gray, "When they dry out, they shrivel up and look like hell. But just add water, and they puff up and germinate."

Besides being of interest to seedless fruit growers, horticulturists can use this technique to produce a clone of a prize-winning specimen. Also these fake seeds will not have to be stored in dark dry places as regular seeds do.

Then there are computers. They will become as much a piece of a farm's standard equipment as tractors. They will have any number of applications from advising the farmer how and where to spread fertilizer to early diagnosis of diseases and suggested treatment. The Farmer's Almanac will be displaced by the hard disc. It is predicted that expanded use of computers will end up cutting costs for the farmers and the consumers. On the downside, expanded use of high-tech equipment, such as computers, will probably accelerate the demise of the small farm. The small farmer does not have the resources for all the capital outlay or the management expertise to adapt quickly. The big farmer will have the edge.

There are some who do not view genetic engineering and the new technology as the best solutions to our food problems. And their numbers are growing. An estimated twenty to forty thousand farms use what is called "alternative farming" techniques.

The alternative farmer concentrates on using nature rather than fighting it. He eschews chemical fertilizers and uses animal wastes instead. He searches for natural pesticides such as pulverized crab shells. (A protein in the shells is used by certain funghi and other microbes to produce an enzyme lethal to nematodes—nematodes feed on roots and destroy an estimated $3 billion worth of crops a year.)

The alternative farmer is interested in finding crops other than wheat and corn, crops that will not cost so much to grow and will not contribute to the erosion of land. They are advocates of polyculture instead of monoculture. Fields would have several crops growing in them instead of only one. Wes Jackson, of the Land Institute, experimented with a polyculture field of Illinois bundleflower, Maximillian sunflower, and Siberian perennial wild rye. The bundleflower has seeds that are thirty-four percent protein. The sunflower's roots discharge a herbicide, and the rye is believed capable of a large seed production. Jackson hopes to find a combination that will imitate a natural prairie. A prairie needs no replanting, tilling, or fertilizing. A prairie, says Jackson, "is running on sunlight instead of fossil fuel. And it's actually accumulating soil instead of losing it through erosion."

While no one, not even its staunchest advocates, thinks there is any chance alternative farming will replace conventional farming in the near future, it will probably attract more and more followers who are disgusted with Alar apples and ethyl dibromide cereal.

While some farmers seek to return to Nature, other trends are moving towards taking food production out of the field and putting it in a factory—not "food processing," but rather the actual growing and fabrication of plant material.

This trend started with hydroponics. Hydroponics is a system of growing crops without soil. Plants are raised in a water solution that contains the nutrients needed for growth. It is a farming factory, all inside, under lights. One company, PhytoFarm, in DeKalb, Illinois, uses conveyor belts that move the growing plants throughout the warehouse until they are harvested near the loading dock.

Gone are the worries about pests and bad weather. Hydroponic farming is in a controlled environment that can produce near-perfect crops year-round.

Tony Mantuano, chef of Chicago's Spiaggia restaurant, has extolled the virtues of hydroponic farming. "The consistency year-round is great. How else could I serve fresh, unbruised spinach during January and February in the Midwest?"

Hydroponics is not for all crops. Best suited are lettuce, spinach, parsley, celery, and various herbs and spices, with strawberries possibly being added to that list.

There is, however, one major disadvantage to hydroponics, one that at times sends the price of ten packs of greens up to almost $20: Electricity. All those lights essential to the process needs an enormous amount of electricity. General Electric had once been very hot on the concept of hydroponics and even opened two plants, one at an old Air Force base in Alaska, a second in Syracuse, New York. But by 1980, after only a few years, GE had closed both plants and sold the technology to Control Data. One observer said the factories "didn't make enough money soon enough."

But countries with cheap hydroelectric power, such as Norway and Sweden, as well as land-poor Japan, are becoming increasingly interested in hydroponics and will undoubtedly be opening more and more farm-factories in the future. Japanese companies are already investing research money and one has come up with a nuance where the nutrients are sprayed over the plants instead of put into the water solution. This method allows the plants to grow even faster—a mature lettuce can be harvested in twenty-eight days as compared to ninety in a greenhouse and 120 the old-fashioned way, out-of-doors.

While hydroponics may sound like the highest-tech end of this concept of "factory farming," something far more extraordinary is in the laboratory: Vat farms.

Picture acres of fast-growing trees, trees being cut and chipped and sent to the factory. The wood chips, composed primarily of sugar polymers, are converted to simple sugars. The sugars get piped into a factory where genetically engineered cells of corn multiply in the sugars and are alchemized into the part of corn that is used for food. So what would be "grown" would be corn kernels and not the whole corn plant.

According to Margaret Mellon of the National Biotechnology Center of the National Wildlife Federation, this would mean "a food source without dependence on plants or animals at all. In tissue culture, we can make orange juice without oranges. It means a divorce between our food supply and the organisms on which it is now based."

It is highly unlikely that this technology will ever be perfected to the point that whole tomatoes, ready to be sliced in a salad, will be plucked from the vat farms. But, tomato "essence" for tomato sauce? Quite possibly. Will it happen? As Malcolm Gladwell wrote in the *Washington Post* in 1989, theorizing about such future food factories is "partly an intellectual exercise," but it also "reflects the conviction of some that given the world's growing demand for food, the escalating costs of farming and the agriculture-threatening consequences of global warming, having more than one way to make a strawberry may be a good idea."

Should man ever get around to colonizing environmentally hostile planets, being able to ship wood chips and strawberry cells to Mars might solve some of the food logistics.

The United States Office of Technology Assessment has made some projections on what the advent of these various new technologies—such as genetic engineering and computer farm management—will mean to the United States farmer. In 1984, the United States produced 7.7 billion bushels of corn. The Office of Technology Assessment has said that if no new technology is used at all, it would be expected that by the year 2000 American farmers would be producing 8.6 billion bushels. However, if new technology is used at the expected rate, that figure jumps to 9.3 billion by the turn of the century. And if new technology gets adopted faster than predicted, we may witness the production of 9.7 billion bushels.

Take another major crop, rice. In 1984, the actual production was 13.7 billion bushels. By 2000, with technology being used at the expected rate, that jumps to 16.3 billion bushels, and 16.9 billion if the technology gets put into place faster. Milk production is expected to go from 135.4 billion pounds to 192.1 billion. And so on down the line.

One thing is for sure. Old MacDonald's farm is being rebuilt from beneath the ground up.

SIX

The Environment: Clean Air, Clean Water, Clean Up

H.G. WELLS told the tale of three sailors shipwrecked on an island with nothing to eat, save a succulent pig that inhabited the isle. Each sailor was obsessed by the pig, or, rather, by one particular part of it that he craved. Even though no one wanted the same portion of the pig, they refused to cooperate with one another in hunting the whole animal. So the sailors found themselves on the verge of starvation when they were finally rescued.

Wells's moral was that men and women of "sufficient intelligence" must realize the simple truth that global problems cannot be resolved in a piecemeal manner based on individual interests. But "if they do not," the British writer warned, "we shall go to an emaciated and miserable end."

That was written in 1928, and for the rest of this century mankind has failed to heed Wells's parable. But in the future we will either learn the lesson or face the miserable end because the problems threatening our environment, threatening the basic life support systems of the planet, have truly become global.

From the emergence of our environmental consciousness on Earth Day, April 22, 1970, through the end of this century, most of the focus on environmental issues has been on specific and often seemingly isolated issues, such as cleaning up toxic waste, finding a place to dispose of our trash, and preserving individual plant and animal species on the verge of extinction.

Each of these battles has been waged on its own field, often without any link to any other environmental issue.

Even environmentalists, for instance, find it hard to explain the link between cleaning up the toxic wastes at Love Canal near Niagara Falls, New York, and the effort to save the Concho water snake from extinction by the construction of the Stacy Dam in western Texas. In fact, the relationships are tenuous because our vision of the environment has been compartmentalized, and that has led to environmental regulation and policy during the last quarter of the twentieth century that was naive at best, disjointed and counterproductive at worst.

That is about to change. The problems of the future are so comprehensive, so global in nature that they will force us to expand both our vision of the environment and the solutions we must use to preserve its balance.

No longer will we simply worry whether an Ohio power plant is dusting its neighbors with soot or debate whether or not a piece of a North Carolina barrier island should be sheltered from development to preserve the natural scenery.

These things will be seen as part of the broader issues of energy, land use, population, and pollution policies that must be resolved to assure the continued well-being of not only an Ohio town or a North Carolina beach, but the nation and world.

The driving force in this transformation is the growing awareness of the interstate, international, and global impacts of man's activities. It began in the mid-1970s with the finding that acid rain, a mild solution of sulfuric and nitric acids, was severely affecting lakes and forests in New York's Adirondack Mountains, New England, and a large swatch of eastern Canada.

Principal sources of the acid rain are the coal-burning electric power plants of the Midwest. These massive power generators burn huge quantities of coal without any pollution-control equipment. As a result, tons of sulfur dioxide and nitrogen oxides are belched from the plants' smokestacks. These gases mix with the moisture in the atmosphere to create acid rain.

Acid rain clearly had an impact on the sensitive Adirondack glacial lakes, on the headwater creeks and streams in Pennsylvania, the ridgeline forests of New Hampshire, and the lakes of Western Ontario. Yet no environmental regulators could do anything about the pollution which was coming from as far as one thousand miles away.

At about the same time that the acid rain issue was emerging, two chemists from the hinterlands of academe published an article that would carry the pollution debate from the interstate and international arenas to the global stage.

In June of 1974, F. Sherwood Rowland and Mario Molina, both researchers at the University of California-Irvine, published a paper in the prestigious British scientific journal *Nature* that stated that a group of widely used man-made chemicals, called chlorofluorocarbons, or CFCs, were building up in the atmosphere and slowly working their way to the stratosphere, some twelve to thirty miles above the Earth's surface.

Rowland and Molina warned that the CFCs could destroy the planet's protective ozone layer, which screens out much of the sun's harmful ultraviolet rays. Ozone is an unstable gas, and the ozone layer is relatively thin and vulnerable. These ultraviolet rays would break down the CFCs, releasing free chlorine. One chlorine molecule, they stated, had the power to set off a chemical chain reaction that could ultimately destroy as many as 100,000 ozone molecules.

Molina and Rowland calculated that between 7 and 13 percent of the Earth's tiny ozone reservoir could be destroyed during the next 100 years—enough to seriously affect life on the planet.

Because their study was primarily a theory of chemistry and physics, it initially promoted a lot of debate and controversy but little action. It did, however, provide us with the first glimpse of the global nature of the environmental problems that were to face us in the future.

The ozone-CFC controversy was still raging when yet another global issue burst onto the scene—the Greenhouse Effect. Like the CFC debate, the implications of the Greenhouse Effect were worldwide.

Quite simply, man's burning of hydrocarbon fuels—from wood to coal to gasoline—continues to fill the atmosphere with carbon dioxide which acts like a pane of glass in a greenhouse, holding in heat from the sun that might otherwise be radiated back out into space.

The more carbon dioxide or other greenhouse gases in the atmosphere, the warmer the atmosphere will get, a number of scientists contend. At the rate we are burning fossil fuels now, by the middle of the next century we will have 55 percent more carbon dioxide in the atmosphere than we do today.

That, according to projections by the National Science Foundation, will lead to a rise in the Earth's average temperature of 3 to 9 degrees Fahrenheit. While that may not sound like much, the difference between today and the last Ice Age, when New York City was buried under a three-mile-thick glacier is only 5 degrees Fahrenheit.

A 5-degree increase in temperatures would, according to studies by NASA's Goddard Institute for Space Studies in New York City, lead to more than a doubling of the chances of a hot summer in the United States, a 25 percent increase in the frequency of droughts, and almost two solid months of above 95 degree temperatures in a New York City summer, a fourfold increase from the present.

Due to the natural expansion of water and the melting of glaciers, the seas would also rise, eroding coastlines and flooding low-lying areas. The warmer water would also feed hurricanes, which would become fiercer and more frequent.

The Greenhouse Effect would touch every continent, and the forces pushing us toward the unstable Greenhouse world are global. Even though the United States is the world's principal energy consumer and carbon-dioxide emitter, the country is incapable of stopping the Greenhouse Effect alone.

Already the United Nations Environmental Program is making the Greenhouse Effect a top international issue, and not only major nations like the U.S. are concerned. Among the most vocal advocates of international action are tiny island nations like Malta and the Maldives. Their concern is understandable. If the oceans were to rise an average of five feet, the Maldives would disappear from the face of the earth.

Growing concerns about air and sea, from acid rain, to ozone, to global warming, to ocean pollution, are stretching our vision of the environment and the problems it faces. Broader vision will dominate the future, and it is in this environmental spectrum that we will ultimately succeed or fail.

Environmental campaigns during the last thirty years were not without some successes, and in many ways these early victories, in pushing new environmental concepts and regulations, helped pave the way for the broader approach of the next century.

The first major environmental issue was that of toxic wastes. As early as 1962, Rachael Carson sounded the alarm in her classic work, *Silent Spring*. While the book focused on the excessive use of pesticides, Carson's fundamental message was that we were introducing deadly man-made chemicals into the environment.

It was not, however, until two incidents in the mid-1970s that the national consciousness and political will were galvanized. The first episode occurred in Elizabeth, New Jersey, at the site of the defunct Chemical Control Corporation. This firm had left tons of unmarked but highly toxic chemicals on the site, which were almost set off by a large fire. The entire city of Elizabeth was at risk, and the state had to clean up the mess.

Then in 1978, the problems at Love Canal, a housing development built atop an old chemical dump in Niagara, New York, turned the nation's attention to the toxic waste issue. When Love Canal residents complained of a host of medical problems, the area was evacuated, the houses were eventually bought by the federal government, and the families were relocated.

The answer to toxic wastes, at least Washington's answer, was two pieces of legislation: the Resource Conservation and Recovery Act, or RCRA, and the Comprehensive Environmental Recovery, Compensation and Liability Act, CERCLA, better known as the Superfund.

Because problems like Chemical Control and Love Canal were caused by the poor handling and disposal of toxic materials, RCRA's goal was to monitor and regulate hazardous materials more tightly. It was a law intended to deal with future handling of toxic materials.

Superfund, on the other hand, was created to deal with the problems of the past—to clean up the messes made during the decades when we neither understood nor cared much about the impacts of chemicals on the environment.

Impacts of the two laws have been mixed. The Superfund has grown from an initial list of roughly 400 sites to be cleaned to nearly 1000 and, according to the EPA, there are 30,000 sites waiting to be assessed. The agency estimates that the list could double by early in the next century and the cost of cleanup could rise to a total of $23 billion.

That, however, may be an optimistic estimate. The Office of Technological Assessment (OTA), the agency that advises Congress on scientific and engineering issues, projects that the list could reach 10,000 sites and the clean-up costs may eventually exceed $100 billion. Just $1.6 billion was appropriated in the original bill. OTA also proposes that the cleanups could take another thirty or forty years to complete.

The reason the list continues to grow is that it takes only a few months' evaluation to get in the Superfund program, but actually cleaning up the toxic wastes has proven to involve years of legal wrangling and engineering and technical difficulties.

That is the bad news. The good news is that a whole new host of technologies is being developed to deal with toxic wastes. Vacuum pumps to suck toxins from the ground, sophisticated resin filters to skim chemical wastes from streams, rivers, and lakes, and new technology to destroy waste materials will make the job of cleaning up sites and handling future hazardous wastes easier, according to OTA.

For example, two technologies OTA found promising are plasma arc reactors and supercritical water. The plasma arc reactors use streams of electrons to break the bonds between molecules. This renders the materials nonhazardous, and it also turns them into a gas that has some fuel value. Tests by the Canadian government found the reactors to be 99.99 percent efficient in destroying hazardous compounds. The process is in the public domain, and there are some tentative efforts to commercialize it.

Supercritical water is heated under pressure to more than three times its normal boiling point (374 degrees Celcius). This superhot liquid turns out to be an excellent solvent for dangerous organic chemical compounds, such as the carcinogenic PCBs (polychlorinated biphenols) or the pesticide DDT. Laboratory tests have found it to be 99.99 percent effective. A system for this technology has already been patented by Modar Incorporated.

Since one of the biggest headaches for the Superfund is figuring out what to do with cleaned-up wastes, these new technologies should help speed the program in the future.

RCRA has already been successful in limiting the production and illegal disposal of toxic residues. The law tightly controls the amount of wastes produced and the disposal of that material. That has been very effective in its own right. But what the law has really done is make toxic wastes very expensive to handle and get rid of, and it has also given communities the right to information about the hazardous chemicals used at their local facilities.

As a result of the law, one federal government study found that the cost of disposal of a ton of hazardous waste rose in the last decade from less than $50 to more than $200. That coupled with the adamant opposition of communities across the nation to being the sites for waste depositories has made toxic wastes both an expensive and politically touchy issue.

In response, industry has begun major efforts to cut back the amount of hazardous waste they produce. The 3M Company, for example, has set a goal to reduce all its hazardous wastes by 90 percent by the year 2000.

Monsanto, Dow, Du Pont, Rohm & Haas, and many other major chemical producers are following suit and instituting "waste minimization" programs. "There is a lot of focus in this area now. . . . It's one of the biggest things going on in the industry," said one official of the Chemical Manufacturers' Association.

Currently, between 225 and 275 million metric tons of hazardous wastes are generated in the United States every year. Waste minimization efforts by industry could cut that figure by between 35 and 50 percent over the next few decades, according to EPA and OTA estimates.

"In many cases, it has gotten to the point where pollution prevention techniques and process modifications are now cheaper to put in place than end-of-the-pipe pollution controls," according to James Lounsbury, director of EPA's waste minimization program.

Improved technologies to handle the waste coupled with the significant reductions in the volumes of hazardous materials will make toxics a relatively well-managed, well-policed activity.

The same can be said of the "solid waste" or garbage problem. As toxics had burst upon the media scene in the 1970s, so trash became big news in the 1980s, as urban communities found it harder and harder to find places to dump their trash. Available landfill locations will practically disappear in the 1990s.

In 1987, the nation followed the almost comic voyage of a barge laden with garbage from Long Island, New York. For 160 days the barge was towed along the coasts of the Atlantic and the Gulf of Mexico. At every port the trash was turned away. Six states and three countries spurned the cargo, and finally the barge returned to home where the trash was incinerated.

Back in 1960, we generated an average of 2.9 pounds of trash daily for every man, woman, and child in the country. By 1988, that figure was up to 3.5 pounds of solid waste, and the number, incredibly, is expanding. In 1988, municipal solid waste for the nation weighed in at 160 million tons, according to the EPA. The agency projects that by the year 2000 that figure will grow another 20 percent to 193 million tons or 6.5 pounds for every person in the nation.

It is not surprising that when the Solid Waste Management Association conducted a poll in 1988, 53 percent of the people questioned thought that garbage disposal was already a national problem. If the trends continue without new solutions the nation could well be buried in its own trash.

Within a matter of decades, however, the problem will largely be solved, for, like toxic waste, the economics of garbage will drive us to find new solutions. Consider these figures: In 1984 Philadelphia was paying about $12 a ton to dump its municipal trash in a landfill; by 1988 the price had quadrupled. That provides a powerful incentive to find another answer to the problem.

The easiest solution is to recycle all valuable portions of the trash stream—the glass, metals, plastics, and paper. Currently, only about 11 percent of American trash is recycled, according to the National Solid Waste Management Association. More than three-quarters of our refuse ends up in landfills.

By contrast, Japan recycles 50 percent of its solid waste, incinerates another 23 percent, and landfills only 27 percent. Obviously there are alternatives.

Nine states—Oregon, Rhode Island, New Jersey, Connecticut, New York, Pennsylvania, Florida, Wisconsin, and Maryland—now have recycling laws. Oregon, which was a pioneer in this field, already recycles about 20 percent of all its trash, and New Jersey has set a 50 percent target in its law.

In just four years after passage in 1983, New York State's law had a dramatic effect. Recycling of plastic milk bottles rose from just 1 percent to 50 percent, aluminum can recycling jumped from 15 to 60 percent, and recycled glass bottles went from 4 to 80 percent.

The list of recycling states will grow in the coming years. Although solid waste disposal has traditionally been the responsibility of local government, by 1988 Congress was considering several bills to help promote recycling. In the future, there will be additional federal help and incentives for recycling.

A major problem with recycling has been the fluctuation in the markets for such materials. All too often there have been gluts of materials, particularly paper, and despite the best efforts of local officials, they have found themselves sending their recyclables to the landfill.

As the availability becomes more stable, the quality more consistent, and government backing for recycled goods stronger, the markets should also stabilize. In addition, the availability of the raw materials will also encourage enterprising companies and individuals to find new uses for the material.

Right now, plastic soda bottles are being transformed into the insulation for sleeping bags, and plastic milk jugs are being used to make a rot-resistant board for waterfront decks. In 1989, Procter & Gamble started marketing its Spic & Span in cleaner bottles made from recycled plastic. The list of products providing new markets for recycled materials will continue to grow.

Recycling will be coupled with "packaging laws" aimed at reducing the waste stream at its source. Packaging accounts for 30 percent of all municipal waste, and one dollar out of every eleven spent for groceries in the United States is spent on packaging.

By cutting the amount of wasteful packaging, the volume of trash will also be cut. In 1988, Berkeley, California, and Suffolk County, New York, banned the use of certain kinds of plastic packages. Other local ordinances will likely follow, for one way of attacking the cost and headache of the mound of garbage facing local officials is to get rid of that 30 percent of packaging. Eventually, the producers of consumer products will take the hint.

One other major alternative has been the use of "trash-to-steam" plants that incinerate trash and use the heat to power steam turbines that produce electricity. There are 122 such plants either operating or slated for construction in 36 states, according to the Institute for Resource Recovery.

By the end of this century, these plants will be handling about 15 percent of our trash flow. By contrast, Japan currently depends upon such incinerators to handle 23 percent of its trash, while West Germany burns 30 percent of its trash.

In the U.S., however, the plants face an uncertain future because of the environmental and economic climate. Both the federal and state governments have moved to tighten air emission standards for such plants and to control the disposal of the ash generated. This ash usually contains traces of potentially toxic heavy metals.

Furthermore, to finance these large plants, which can cost as much as $300 million, counties and municipalities must sell development bonds. The economics of the plant depends upon the bond rates, the tipping fee it receives for handling the trash, and the price for which it can sell its electricity.

It is a difficult equation. A plant in Tampa, Florida, for example, opened in 1985 and in 1986 ran a $6 million deficit. In 1987, it ran $1.5 million in the red. A small plant in Tuscaloosa, Alabama, averaged $1 million in losses for each of its first three years of operation.

The future of trash-to-steam in the United States will depend upon the environmental and economic performance of the 122 plants now in development or operation. It is obviously a technology that works. But it will take another decade to determine whether it will become a major component in handling our trash.

Nevertheless, the combination of recycling, package reduction, trash-to-steam, better controls on landfills will all gradually reduce the critical nature of the nation's trash problem.

And so toxic waste and solid waste—two of the biggest headline generators in the last few years—will recede into the humdrum, business-as-usual world. Of course, figuring out how to clean up after ourselves does not rank as any great achievement when we realize it took more than a generation to come up with some answers.

Unfortunately, there are a lot more failures and half-successes in our environmental record. The Clean Air Act, for example, has had a mixed track record. In the ten years between 1978 and 1987, the amount of sulfur dioxide, soot, organic chemicals, carbon monoxide, and nitrogen oxide—all serious pollutants—decreased between 8 and 25 percent. Lead in the air decreased 97 percent after the EPA ordered the cutback of leaded gasoline in the mid-1970s.

Impacts of the act on a city like Philadelphia are marked. In 1966 automobiles emitted 458,000 tons of organic chemical fumes into the air. By the late 1980s, it was down to 273,000 tons, and that rollback was made even as the number of cars in the city rose. The amount of sulfur dioxide in the air was cut almost 90 percent during the same period.

Despite this progress, the air in Philadelphia still remains smoggy during the summer, in violation of the Clean Air Act, and now it is also evident that industries in the area discharge more than 29,600 tons of toxic chemical fumes into the air annually. More than one hundred other urban areas across the nation are in the same straits.

The Clean Water Act has had similar mixed results, as has the Endangered Species Act, which was designed to ensure that plants and animals on the verge of extinction due to man's activities would be protected. There is a backlog of more than thirty-five hundred species just waiting to get on the list, and eighty of those plants and animals are believed to have become extinct while waiting.

Even getting on the list is no guarantee of protection. The Attwater prairie chicken is on the list, but its habitat has shrunk 80 percent. In fact, the struggle over habitat, or land, is at the core of the failure of this legislation. Since the Pilgrims first landed at Plymouth Rock, it is estimated that the country has lost more than five hundred species of plants and animals. We have also lost 50 percent of our forests and wetlands.

"Whenever the act gets in the way of somebody's pet project, they always find a way to get around it," said John Fitzgerald, an attorney for the Defenders of Wildlife, a national environmental organization.

Of course, even the beleaguered Endangered Species Act has had its successes. There are still bald eagles, brown pelicans, and

American alligators because of the protection the wildlife management act afforded.

But how does one measure these victories—a bird saved from extinction, a little less of one particular pollutant floating in the air, a reduction in toxic wastes coming from a chemical plant? Quite simply, they really can't be totally assessed. The victories are limited, and the failures are often linked to larger issues which we fail to perceive or refuse to address. The activities we have undertaken in this scattershot approach will continue, but they will have to be drawn into a broader context. In doing so, ineffective laws, like the Endangered Species Act, may take on new meaning and added clout.

If that does not happen, H. G. Wells' "emaciated and miserable end" may be lurking in our future, for the problems we will face in the next century will be far more complicated. They will be problems that involve man's basic relationships with the air, the land, and the sea, and they require far more comprehensive solutions.

For nearly a century, air pollution had been regarded as basically a local phenomenon, but in the early 1970s, the work of a handful of scientists began to change that perception and also our perception of how we must manage our environment.

At the University of California-Irvine, Rowland and Molina had done their work on CFCs and the ozone layer, and at Washington University, Eugene Likens had published the seminal work on acid rain.

In both cases, the scientists found that their scholarly work had pitted them against major economic and industrial interests. Molina and Rowland's findings that CFCs were slowly working their way up to the stratosphere, where they could break down and destroy the natural ozone layer, was not welcome news to the American chemical industry which was producing almost 1 million tons of CFCs annually, worth about $750 million. The value of goods and services directly dependent upon the chemicals was estimated at $28 billion.

Likens' research, which showed that the emissions from coal-burning plants created acid rain that could end up falling hundreds of miles from the plant, ran afoul of the electric utility industry, which relies on coal to generate much of its electricity. Coal producers in both the East and the West opposed Likens' findings.

These were big dollar interests, and the initial studies were ferociously attacked. But over the years, the findings have withstood both political and scientific scrutiny, and the issue no longer is whether the premises are true, but rather what the impact may be.

Those impacts, however, are often subtle, and it is only slowly that we have begun to understand the future risks they may pose. In the case of acid rain, for example, Likens' initial studies preceded by several years any documentable effects from the precipitation.

However, by the late 1970s clues to the effect of these acids in the environment finally began to emerge. The most publicized was the ecological destruction of some two hundred lakes in the Adirondack Mountains. The acid rain itself did not destroy the lakes, but rather it leached, or flushed, aluminum from the soil around the lakes. The aluminum, in turn, poisoned aquatic life.

Acidity is measured by its pH, which is a count of the hydrogen molecules in a solution. The lower the number, the more acidic it is. A neutral solution, which has no acidity or alkalinity, has a pH of 7. Battery acid has a pH of 1.

Natural rain is slightly acidic with a pH of just above 6. The rain falling on Pennsylvania and parts of New England has an average pH of 4.2. That is one hundred times more acidic than normal rain. Trout cannot live in water with a pH lower than 4.7.

The Adirondack lakes are, however, only the tip of the iceberg. Lakes, streams, forests, and many man-made structures also face serious disturbance from acid rain. So far, these threats have been masked by nature's ability to absorb the acid.

There are a number of naturally occurring substances, such as calcium carbonate, that neutralize the acid rain. While New York State had little such material in its mountain soils, neighboring Pennsylvania had considerably more, and so the impact of acid rain has not been as pronounced. But that carbonate will only last so long.

The Pennsylvania Fish Commission, which monitors lake and stream quality, has stated that waterways in the state are becoming increasingly sensitive. So far, the commission has stopped stocking 816 streams because of acidity. It calculates that another 1,366 streams are now "vulnerable to acid rain."

That warning can be repeated for state after state. The EPA conducted a survey of lakes in the Northeastern United States in 1986 and found that 49 percent of the nearly one thousand lakes sampled were already "acid sensitive."

Using the EPA data, the Natural Resources Defense Council, a prominent environmental organization, has calculated that there are probably sixteen thousand acidic or acid-sensitive lakes in the twenty-three states the EPA has done sampling.

After years of scientific research and debate, it is also now clear that acid rain plays a role in the distress and decline some forest areas are experiencing.

Research done by the U.S. Forest Service's Forest Response Program has found that acid rain may damage the needles of red spruce, impairing the trees' ability to photosynthesize. It has also found that acid rain may also leachate minerals, like aluminum, that can damage a tree's roots.

The effects of acid rain are being complicated by those of smog pollution, which may damage forest leaf cover. City smog used to be considered a local, urban problem, but studies now indicate that, like acid rain, it can travel for hundreds of miles.

Acid rain is also washing away much of our heritage. The National Park Service estimates that in the region covering Pennsylvania, Maryland, Virginia, New York, and West Virginia more than five hundred buildings and twelve hundred monuments are being eaten away by acid rain.

Henry Magaziner, a former parks service architect, said that when he joined the service in the early 1970s, the two lions that guard the entrance to the Merchant Exchange, a colonial building in Philadelphia, had well-defined hair and sharp eyes. "Now," he testified before a Pennsylvania state legislative committee, in 1987, "they are almost formless."

The message is clear: While we avoided widespread acid rain problems in the twentieth century, we will see our lakes, streams, and buildings severely affected in the future, unless steps are taken to avert the problem.

On average, about 21 million tons of sulfur dioxide are spewed into the air in the U.S. annually. More than 14 million tons come from electric utilities and another 5.2 million tons are from industrial fuel and process activities.

The Bush administration and Congress have agreed on a plan to reduce those emissions by ten million tons by the year 2000. Understandably the electric utility industry is not happy about this, and the Edison Electric Institute, the industry's research organ, predicts that the plan will cost electricity ratepayers as much as $7.1 billion a year over the next twenty years.

Environmental organizations contend that the industry figures are overstated and support the 50 percent cut in emissions. But that may not be enough to deal with the whole problem. Scientific studies in the early part of the next century will be key in determining if our lakes and forest stabilize and regenerate. If those studies indicate that the trend toward degradation is continuing, then even bigger reductions may be necessary. That issue will surely create more hot debate in Congress.

At least this is a problem over which we have some control. West Germany, which is also suffering the effects of acid rain and air pollution on its Black Forest, is stymied in dealing with the problem since much of the pollution comes from Eastern Europe.

For the United States, however, of all the air pollution problems facing us in the twenty-first century, acid rain may be the easiest with which to deal. The others—stratospheric ozone depletion and global warming—are much more complex.

Back in the early 1970s, CFCs were seen as part of the solution, not a problem. These nontoxic, nonflammable compounds were used in a wide range of products and industrial processes. While perhaps the best known use was as a propellant in hairspray, underarm deodorant, and furniture polish, the compounds were also used to cool cars, refrigerators, homes, and shopping centers; to clean computer chips; to insulate homes and offices; to create softer cushions, carpet pads, and car seats; and to sterilize hospital equipment.

Terrestrial life on this planet owes its existence to the ozone layer, which filters out harmful ultraviolet light. "Without ozone," explains Brian Toon, a scientist at the National Aeronautics and Space Administration Ames Research Center in Moffett Field, California, "we'd all have to go back to living under the water."

Rowland and Molina originally estimated that CFCs could cause the loss of 7 to 13 percent of the planet's ozone shield. That, unfortunately, was before the ozone hole over the Antarctic was discovered in 1985. The hole had an ozone depletion of more than 40 percent, a loss exacerbated by the isolated nature of the continent. It turned out that there was an even more potent form of atmospheric chemistry than Rowland and Molina had considered, and this chemical reaction could destroy even more of the delicate, gaseous layer over certain parts of the globe.

But scientists, such as Harvard University's Michael McElroy, now voice concern that this same type of chemical reaction could be happening other places in the world. An international expedition to the Arctic Circle in 1989 found the presence of CFCs in the atmosphere and concluded the Arctic, too, was "primed" for a major loss of ozone.

The poles, however, are not the only areas at risk. McElroy, who is the Abbott Lawrence Rotch Professor of Atmospheric Science at Harvard, has warned that the conditions over the tropics also may be conducive to large-scale losses of ozone.

In March 1988, a team of one hundred scientists, known as the Ozone Trends Panel, reported that ozone losses were occurring not just at Antarctica, but throughout the world. It found that from 1969 to 1986 average ozone levels in the Northern Hemisphere had declined by 1.7 to 3 percent.

The most heavily populated regions of Europe, the Soviet Union, and North America had suffered a year-round loss of 3 percent and a depletion of 4.7 percent during the winter, according to the panel.

Making matters worse, McElroy has also calculated that because CFCs are very durable chemicals and so much has been released into the atmosphere in the last few decades, we face a problem of ozone depletion throughout the twenty-first century, even if dramatic CFC cutbacks are made now.

What does this presage? The ultraviolet rays penetrate into living cells and impair their functioning and reproduction. This will manifest itself in a variety of ways.

Every one percent loss of ozone, scientists estimate, will result in about a five percent increase in skin cancer cases, resulting in as many as 15 million extra cases. That means that at least one child in every first grade class in the next century will, during his or her lifetime, get skin cancer because of exposure to the increased ultraviolet radiation.

Ozone depletion might cause up to 2.8 million cataracts and millions of cases of immune disorders in the United States during the next century, scientists predict. Ultraviolet light may also produce mutations of rare virus types.

An increase in ultraviolet light could also adversely affect the planet's plant life both on land and in the oceans, reducing yields of vital crops and impairing entire ecosystems.

Alan Teramura, a botanist at the University of Maryland, has found that ultraviolet rays damage cells and tissue in about two-thirds of the two hundred species of plants he has tested. He found that a simulated ozone loss of 25 percent reduced the yield of one important soybean species by as much as a quarter. Yields of peas, beans, squash, cabbage, rice, and wheat were all reduced when exposed to increased ultraviolet light.

Studies also show that a 25 percent loss of ozone reduces the productivity of phytoplankton by 35 percent. Phytoplankton, the tiny plants that float near the ocean's surface, are the foundation of the entire ocean food chain. The loss of phytoplankton would eventually impact our key fisheries.

At this point, however, it does not appear that we will lose as much as 25 percent of our ozone layer in the coming decades, and the impacts on plants and ecosystems may be harder to document.

Nevertheless, we are entering a new world, one in which we will pay much closer attention to both our own and our world's exposure to ultraviolet light. Tanning will doubtless be classed with smoking cigarettes as a life-threatening habit.

And protecting the ozone layer will take on greater significance as a policy issue. The discovery of the Antarctic ozone hole was enough to galvanize the nations of the world into formulating the 1987 Montreal Protocol. Under this accord, forty nations have undertaken to reduce their CFC production by 50 percent by 1998.

But that now appears to be insufficient to assure the integrity of the ozone layer. Susan Solomon, a top scientist with the federal National Oceanographic and Atmospheric Administration, has warned that a 50 percent reduction will still result in doubling the amount of CFCs in the atmosphere during the next century. She has calculated that at least a 95 percent reduction in CFCs is needed to return the ozone layer to stability by the end of the next one hundred years.

In 1985, there were about 2.7 parts per million (ppms) of chlorine (the actual ozone destroying element in CFCs) in the atmosphere. Even with the Montreal Protocol, the chlorine level will rise to about 6 ppms in 2025, 11 ppms in 2065, and 14 ppms in 2100, according to McElroy's calculations.

Clearly, more has to be done. In early 1989, the European Economic Community decided it would attempt to phase out all CFCs by the year 2000. Later that year, an international conference in Helsinki also called for a ban on CFCs by the turn of the century.

The Bush administration has also committed to a prompt phase-out of CFCs as soon as viable alternative chemicals are available. Those substitutes are quickly being developed. Both Du Pont, which is the largest producer of CFCs in the world, and Imperial Chemical Industries (ICI) have announced the development of "ozone-friendly" CFC substitutes.

ICI, for example, plans to begin production of an alternative fluorocarbon for CFC-12, which is used in automobile air conditioners and domestic refrigerators, by the early 1990s. Dr. B. H. Lochtenberg, ICI's chairman, said he hoped with this move his company "will be favorably positioned to be the global leader in commercial production of 'ozone-friendly' CFC replacements."

The ozone issue in many ways has provided a model for future international political and industrial action in these global issues. That model will be severely tested on the other great environmental problem of the next century—global warming.

There are some countries, however, such as China and India, that have not moved as quickly on CFCs. They face the dilemma of having invested in the technology for producing CFCs and are now depending upon them for their refrigeration. These nations find switching to new technologies and more expensive substitutes a luxury they cannot afford.

It may take the aid of more technologically developed nations, such as the United States and the members of the European Economic Community, to deal with this problem, and it may require substantial transfers of technology and monetary aid.

As a result, it remains unclear how quickly these CFC compounds will be taken off the market worldwide and therefore how grave our ozone depletion will be in the 21st century.

While acid rain and ozone depletion are both complex, large-scale problems, they appear simple when compared to what will be the ultimate environmental problem of the future—the Greenhouse Effect.

The Earth has thrived on the Greenhouse Effect for millennia; the problem in the next century will be that we may be getting too much of a good thing.

The sun bathes the Earth twenty-four hours a day with its heat, but without our atmosphere to act as a layer of insulation, the planet's surface would only warm to about 0 degrees Fahrenheit. The Earth would be a frigid ball. Fortunately, certain gases in the atmosphere—primarily water vapor, carbon dioxide, and methane—trap some of the heat that would otherwise escape the planet and raise the average surface temperature to about 60 degrees F.

This is the so-called Greenhouse Effect. The problem is that the more greenhouse gases the atmosphere contains, the more heat is trapped. Venus, for example, has an atmosphere composed of 97 percent carbon dioxide. The temperature of its atmosphere is about 900 degrees F.

In 1957, Roger Revelle, a researcher at the Scripps Oceano-graphic Institute, who was among the first to sound the alarm over the Greenhouse Effect, contended that the steady burning of fossil fuels, which release carbon dioxide, would lead to a steady buildup of the gas in the atmosphere and would elevate the temperature of the planet.

Back then, Revelle's theory had no data to back it up. In 1958, as part of the International Geophysical Year, a worldwide scientific research effort, carbon dioxide monitoring was begun, and within a decade Revelle's theory was confirmed.

Carbon dioxide, the main greenhouse gas, has increased by 25 percent, from 280 parts per million a century ago to 355 ppms in 1989, and its concentrations are increasing by slightly more than 0.3 percent each year.

Carbon dioxide is given off mainly by burning coal, oil, natural gas, and other fossil fuels in electric power plants, cars, and industry.

Unfortunately, those are not the only sources. The burning of forty million acres of tropical forests annually is also spewing 1.7 billion tons of carbon into the atmosphere. The destruction and clearing of the forests present a double whammy because plants absorb carbon dioxide and release oxygen through photosynthesis.

As researchers became more attuned to the issue, they found that still other greenhouse gases were mounting in the atmosphere. Methane, which is twenty-five times as effective in trapping heat as carbon dioxide, is increasing at an annual rate of about 1 percent. Methane is given off by termites, landfills, coal mining, cultivating rice, burning fossil fuels, and leakage of natural gas. It is also given off by the world's 1.2 billion cattle.

Nitrous oxide, or laughing gas, is about 250 times as effective at trapping heat as carbon dioxide. This chemical is given off by burning fossil fuels that contain nitrogen and by spreading nitrogen-based fertilizers on fields. It is increasing at about 0.2 percent a year.

It turns out that CFCs are not only bad for the ozone layer, but that they also contribute to the Greenhouse Effect. They are about 10,000 times as effective as carbon dioxide as a greenhouse gas, and they have been increasing at about 5 percent a year.

There are no laboratory experiments to be done to divine the fate of the Earth. Our best tools have been fairly complicated computer models of the globe—its oceans, atmosphere, and land. By simulating various changes in atmospheric levels of greenhouse gases in the models, the researchers have been able to develop different scenarios for the future.

There are, however, three serious drawbacks inherent in our present models. One is our incomplete understanding of some of the Earth's basic processes—cloud formation, ocean currents, and evaporation. This means that the computer-generated scenarios are intrinsically crude.

The second drawback is the spottiness of the basic data we feed into the models. For example, there are whole chunks of Asia and Africa for which we have very little temperature and rainfall data.

Nevertheless, Stephen Schneider, the head of the modeling group at the National Center for Atmospheric Research in Boulder, Colorado, maintains that the models "are in the ball park." The mathematical models have been pretty accurate in replicating the climate, as we understand it, of previous epochs on the earth and on other planets. This, Schneider contends, shows that they are useful tools.

The third limitation of predicting the future from computer models lies in the activities of man. We can change the future by changing our behavior to moderate the production of greenhouse gases. However, it is clear from looking at the list of the sources of greenhouse gases that doing so will not be easy.

It seems as if these gases are an integral part of so many essential activities—energy production, transportation, agriculture, manufacturing, and development. So, in trying to look at the future, let us assume that there are not many changes in our economic behavior during the next few decades.

The three principal climate modeling groups in the United States are those at the National Center for Atmospheric Research, the Goddard Institute for Space Studies, and the National Oceanographic and Atmospheric Administration's General Fluid Dynamics Laboratory in Princeton, New Jersey.

All these groups roughly agree that if the amount of carbon dioxide reaches about 560 parts per million in the atmosphere—something which is expected to happen around the middle of the next century—the Earth will warm between 1 and 5 degrees C.

There are some scientists that say the change will be barely perceptible; there are others, however, that contend that the planet's average temperature may rise by more than 9 degrees C over the next century. Such a rise, most researchers agree, would be catastrophic.

Sticking with the best available projections, what can we expect? All in all, the world's climate will change dramatically. According to a Goddard Institute scenario:

- Droughts will increase substantially over the next few decades and the centers of continents will become drier and drier. The chance of drought in the United States will have increased 25 percent by 2020 and 45 percent by 2050.
- Hurricanes will become fiercer and more frequent. These storms are created and feed off warm water. There will be more warm water and consequently more hurricanes. Kerry Emanuel, a researcher at the Massachusetts Institute of Technology, says there may even be "hypercanes" with winds 50 percent stronger than anything we have seen this century.

- Temperatures will rise in general. There will be, for example, many more days with temperatures above 90 degrees F. New York City will have almost two months of days 95 degrees F. in an average summer by 2030, a fourfold increase over today.
- Sea levels will rise as the earth warms. Part of this rise will be from the expansion of the water in the oceans and part from the melting of glaciers. Over the course of the next century, the sea will rise as much as three feet.

Three feet may not sound like much, but every one-foot rise takes a 200-foot-deep bite out of the coast. A three-foot rise adds up to a 600-foot loss of land along the coast.

"The lowlands along the world's seas will be the areas most vulnerable. They include the deltaic barrier island, atoll, and marshy coastlines," Pier Vellinga, of the Dutch Environmental Ministry, and Stephen Leatherman, of the University of Maryland's Laboratory for Coastal Research, wrote in the *Climate Change Journal.*

The authors went on to say that a sea-level rise, exacerbated by storm surges, will threaten life in the Third World and will require developed countries to expend substantial funds to protect their coasts.

James Titus, a policy analyst with the EPA, has calculated that a three-foot rise in the oceans would inundate between five thousand and ten thousand square miles of dry land in the United States. The cost of protecting the land and the structures on it, Titus has estimated, could turn into hundreds of millions of dollars.

These changes in climate will create a host of new issues and new crises for society. Much of our infrastructure, such as reservoirs, built and positioned on old rain patterns, will be obsolete, as will many sea walls, storm sewer systems, and harbors.

Agriculture in the Midwest will have to be retooled, using new crops or new plants that can grow with more heat and less water, or else the region may be abandoned to prairie.

Warmer summers will worsen city air pollution and bake urban areas. "There will be times that the inner city will be absolutely unbearable," predicts University of Delaware geographer Lawrence Kalkstein. He says that in the future we may need air-conditioned "cool" shelters for the poor, the way we have winter shelters for the homeless today.

Our use of coastal land, which will be vulnerable to rising oceans, storm surges, and big hurricanes will have to be reevaluated. How much, for instance, will we be willing to spend continually in disaster aid so people can save their summer homes at the shore?

Flooding lowlands in the Third World and severe drought will lead to starvation and massive refugee problems. Indeed, for all the problems a warmer world could create for the industrialized world, even a modest temperature increase could wreak havoc in the less-developed nations.

"The capacity of poorer nations to adapt to coming change, and minimize their own contributions to it, is sharply constrained by their limited resources, by their debt problems, and by the need to develop their economies," according to a statement issued at the World Conference on Climate Change, held in Cairo in late 1989.

The real scenario, of course, could be better or worse than the one described above depending upon the accuracy of the models and the amount of greenhouse gases that rise skyward.

One of the confounding elements of the greenhouse problem is that if we wait for it to manifest itself, it will be too late to do anything about it. So what can be done now?

First, the global warming problem is forcing policy makers around the world to think more expansively about the environment. By 1989, bills were being filed in the U.S. Congress with titles like "The Global Environmental Protection Act" and the "World Environment Policy Act." No such bill has yet passed, but it is clear that people are beginning to think in these terms, and very soon our vision of the future will be clearer.

There are already ambitious research programs being mapped out to study clouds and oceans and improve our basic climatic data. In the next decade models will be more sophisticated.

Acid rain and CFC problems now underline the broadening

problems and the need for more coordinated action. The Greenhouse Effect takes us a step further.

For example, the planting of more trees and the better management of the world's forests become increasingly important as a potential damper on the rising load of carbon dioxide. Such programs also reinforce efforts to preserve the habitats of endangered species.

Similarly, a wiser use of coastal lands and more limited coastal development, in the face of a rising sea, helps to protect wetlands, which have been disappearing from the United States at an alarming rate. The U.S. has already lost half its 215 million acres of wetlands, much of it along the coasts. A good coastal-use policy would also help us maintain the quality of coastal and estuary waters.

John Topping, the director of the Washington-based Climate Institute, said that global warming has become "a focal point" for all environmental issues and policies for the coming century.

Perhaps the most difficult problem to solve will be that of energy consumption, which is the driving force behind both our economy and the future greenhouse world.

Each year we expel twice as much carbon dioxide from fossil fuels into the air as the Earth can naturally absorb. The rest lingers and builds up in the air. Now, the carbon dioxide is being joined by other even more efficient greenhouse gases.

It probably comes as no surprise that the largest contributor of this carbon dioxide is none other than the United States. With just five percent of the world's population we create 20 percent of all the carbon dioxide emissions in the world.

About a third of that carbon dioxide comes from electric utility plants and about another third from transportation. Texas, California, Ohio, and Pennsylvania are the four largest generators of the gas.

Other countries, in their effort to follow the model of industrial prosperity, are catching up. For instance, South Korea emitted just 0.7 tons of carbon dioxide back in 1950, according to an EPA analysis. By 1970, that was up to 10.2 tons and in 1987 it had reached 45.5 tons, all a sign of its booming economy. Still, South Korea is just the twentieth largest carbon dioxide generator in the world. The U.S., at the top of the list, is followed by the USSR, China, Japan, and West Germany.

How can South Korea or Brazil be denied their chance to prosper? How can the United States retool its energy consumption without laying an undue burden on its industries or jeopardizing its economy? These questions must be answered if we are going to cope with the greenhouse issue. Without new approaches, efforts to save and increase forests, preserve coasts, and rebuild our infrastructure may not be enough to protect us from the harsher effects of global warming.

SEVEN

Energy for a New Era

THE EVIDENCE must be clear to anyone, except perhaps apologists for the oil and coal industries, that we must loosen the death grip that fossil fuels have on the world economy. We must develop alternative energy sources.

The United States alone produces 1.4 billion metric tons of carbon each year. That is five tons per man, woman, and child.

Yet, in recent years there has been no rush to abandon fossil fuels for cleaner, less harmful energy sources. It is not because these sources do not exist. They are there—renewable, sustainable—but unfortunately more expensive. And therein lies the major drawback to alternative energy sources.

The United States could reduce its air pollution and acid rain, and contribute to the slowing of global warming, by using renewable sources of energy such as the wind, sun, and oceans. But the incentive is not yet there, not with the price of gasoline lower today— with inflation taken into account—than it was in 1974.

Gas and oil are cheap. Alternative energy, in comparison, is not. It is highly unlikely that we will foresake our attachment to fossil fuels until the cost of using these other energy sources drops and the cost of not using them rises.

111

Lowering the price of alternative energies takes money for research, and, regrettably, that was sorely lacking in the eighties. The Reagan years saw an 82 percent decline in research and development funding in these alternative fields. Private industry had no incentive to allocate money because the price of oil was low and tax credits had been eliminated.

This is not to say no progress has been made. Progress has been made, and certain technologies hold great promise for the future.

Perhaps the alternative energy source that received the most publicity and made the greatest impression on the public during the Arab oil crisis was active solar—or solar thermal—power. Solar collectors sprouted up on buildings and installations all over the country. Great predictions were made about solar power. Hopes were raised that it would be the cheap, clean energy source of the future.

It seemed like such a logical solution—use the sun's heat to produce electricity. In this type of solar technology, collectors with mirrored surfaces concentrate the sun's rays onto receivers. The receivers heat liquids, which in turn drive turbine generators. Homeowners who installed solar hot water systems witnessed payback periods of five to seven years derived from utility savings.

There are different types of receivers. The Israelis are using solar ponds—bodies of water, manmade or otherwise—that capture the sun's heat. Another type is called the central receiver. It is placed high above a field of mirrors, the angles of which can be changed as the sun moves. However, the type of solar collector that has shown the most promise is the parabolic trough which has been put into commercial use by Luz International Limited in California's Mojave Desert. There, seven commercially viable plants with 200-megawatt capacities were installed for Southern California Edison.

Estimates project that within the next several decades advances in design, materials, and conversion and storage systems will lower the costs of solar thermal energy far enough to make it a competitive alternative.

But it is another type of solar energy that Renew America, a nonprofit energy group, predicts "will become one of the world's leading sources of energy." With this technology, the sun's rays are turned directly into electricity. Light is composed of both waves and particles. When the particles hit the surface of photovoltaic cells, which are usually comprised of silicon, an electron is displaced. Electrons are then "captured" and flow into the electronic currents forming circuits.

Although the principles behind solar energy were first discovered in 1839, it was not until 1954 that D. M. Chapin, C. S. Fuller, and G. L. Pearson, working at Bell Laboratories, invented the solar battery—the first efficient means of directly converting a significant amount of the sun's energy into electricity.

The cost of producing electricity with those early silicon cells was extremely high—hundreds of dollars for one kilowatt hour (a kilowatt hour being the amount of electricity needed to burn ten 100-watt bulbs for one hour). By 1965 the cost had dropped to $40 per kilowatt hour; in 1973, $12.50; and by the end of the eighties, it was down to between thirty and forty cents. But that still did not make photovoltaics competitive. In Connecticut, for example, it was costing consumers only ten cents per kilowatt hour for electricity generated by conventional sources.

Still, several developments, some still in the laboratory, have made scientists and advocates of solar energy optimistic. For one thing, cells have become significantly more efficient. The early cells could only convert 6 percent of the sun's energy into electricity. By 1989 that capability had doubled. A scientist in Palo Alto was even able to raise the efficiency level to a 28 percent conversion.

Another selling point for photovoltaics is their versatility. They can be used to power Mickey Mouse watches, homes, or, when combined into large modules, full-scale utility plants. This gives a power company the option of building plants to meet the current needs and adding more output later. With conventional fuel sources, to make construction cost-efficient, billions of dollars must be invested in plants that are larger than necessary in anticipation of future needs.

Reliability of photovoltaics has increased, as well. Two California power plants that have used photovoltaic cells for years reported few problems. Such reports are creating increased interest from other utilities. Pacific Gas and Electric is building the world's largest photovoltaic project, which it hopes will serve as a demonstration model for other companies.

This progress has led Edgar A. DeMeo, manager of solar programs at the Electric Power Research Institute in Palo Alto to say, "It's conceivable by the mid-90s . . . we will begin to see a viable power option here for large-scale utility use."

The Department of Energy has indicated it expects utility companies to begin construction of solar plants in the next decade. The Solar Energy Research Institute expects that within forty or fifty years half of all American electricity needs will be met by photovoltaic plants.

Even the Saudis, who have at least one hundred years of oil reserves, are getting into photovoltaics. They have built a $26 million solar facility in the desert thirty-two miles northeast of their capital, Riyadh. The facility consists of 160 photovoltaic generators with 251 solar cells, capable of generating 350 kilowatts of direct current daily. This plant supplies the needs of three nearby villages of four thousand residents.

Although it has proved a costly project, Fahad Huraib, a director at the King Abdul-Aziz center for Science and Technology, said in 1989, "We knew it would be expensive at first, but most of the money went into research. It helped in promoting this type of technology. To develop an energy source takes from fifty to one hundred years."

While solar power cells and peripatetic electrons cast a high-technology futuristic aura, wind power conjures up images of picturesque windmills dotting the Dutch countryside.

A trip through Altamont Pass, near Livermore, California, would quickly dispel that notion. There, fifty square acres of ranchland are covered with seventy-five hundred state-of-the-art wind turbines, generating enough electricity to light and heat sixty thousand homes. Half of the turbines are connected to computers that monitor the wind and allow for the most efficient use of the machines.

By the end of the eighties, wind turbines in California—along Altamont, Tehachapi, and San Gorgonio Passes—and in Hawaii were producing 2 billion kilowatt-hours of electricity, or enough to power the city of San Francisco. The cost of wind energy has dropped from 9 to 7 cents per kilowatt hour, putting it at about the same level as conventional energy sources.

But the inherent problem with wind power has always been the wind itself. It is fickle. Even at the blustery Altamont Press, the turbine blades only spin, on a yearly average, 20 percent of the time.

Besides the capriciousness of the wind, there were difficulties with the technology. Many of the early generation of wind turbines were neither efficient nor reliable. It was thought at one point that bigger would be better, and that was where research money was directed. Turbines got bigger and bigger until finally they were built with blades that swept across one hundred yards of air. Unfortunately, they were dinosaurs that died under their own weight—their shafts kept cracking. Pacific Gas & Electric had one such failed giant in Solano County, California; the company finally dynamited it out of existence.

Learning from prior operating experiences, the trend is now toward smaller machines with a 200–500 kilowatt range. The goal is to develop a new generation of turbines that are cheaper and more reliable. During the next decades it is expected that energy companies will be erecting turbines in the East and in the Northern Plains, where the wind can be depended on to blow steadily for the entire winter.

Wind turbine expert Robert R. Lynette estimates that by the year 2000, advances in the field could bring a generating capacity of between 6,100 and 31,000 megawatts in the West. Projecting further, he sees the possibility that in the early part of the next century wind power could account for 10 percent of all American electrical needs.

According to Mark Haller of Sea West Energy Group, an Altamont Pass operator, foreign countries are showing great interest in wind power. "Not too many weeks after Chernobyl, these hills were crawling with Eastern-bloc types."

Ocean power is probably the least advanced of renewable energy technologies. Still, the movements of the tides and waves and the sun's heat captured on the ocean's surface all offer tempting potential. Harness their action, redirect the heat, and power can be generated.

The French have constructed a commercial 240-megawatt tidal facility at La Rance, France. Chinese developers have a 10-megawatt facility in the Zhejang province, and the Canadians have a plant in the Bay of Fundy.

Wave energy is being used on a small scale to power lighthouses and navigational buoys. Ocean thermal power, on the other hand, is still in the research and development stage, although it has been worked on for the last fifty years.

Although none of these technologies would appear to be major players in the next decade, more countries are likely to allocate research funds to them as further demands are made to find fossil fuel alternatives. It will be only then that there will be any likelihood of tapping the ocean's potential to generate enormous power.

There is one renewable energy source that, although it already supplies 10 to 14 percent of the country's electricity, has fallen out of favor. That is hydropower, and its expansion has been halted by environmentalists.

Dams often damage aquatic life and ruin recreational areas. The National Wild and Scenic Rivers Act of 1968 barred thousands of miles of rivers from hydropower use. In 1988, the Northwest Power Planning Council put forty-four thousand miles of river on its protected list. Advocates will have to show that expanded hydropower usage does not have its own negative environmental impact if it is to grow in the future.

Geothermal power—harnessing the heat trapped within the earth—has run into environmental opponents, as well. It has grown from 900-megawatts of capacity in 1968 to almost 2700 in 1989, with most of the plants located in California. Most of the worldwide geothermal facilities are in volcanic regions, and that is why one major controversy has sprung up.

A geothermal plant has been on the drawing boards for years on the big island of Hawaii. The first notes of discord were sounded by native Hawaiians who protested what they viewed as the defilement of the volcano, which had religious significance to them. Then environmentalists joined the fray, arguing that America's one large tropical rain forest—through which roads have to be cut to reach the mountain—should not be destroyed to provide tourists with more power for air conditioning. The lesson is that no matter how nonthreatening a power source might seem, it must not be approached with blinders on. Side effects and the ancillary costs have to be considered and weighed.

That is why nuclear energy has proved too costly to pursue. Just looking at the straightforward construction costs, nuclear is far more expensive than fossil fuel and as much as 50 percent higher than some of the renewables. The average construction cost of a nuclear plant is more than $3 per kilowatt and has a generating cost of 12 cents. Those figures do not include the high costs of waste disposal. Despite all the French claims, it is not a safe technology nor is it likely that an "inherently safe" reactor will ever be designed. *Nucleonics,* a nuclear industry publication, has stated that "experts are flatly unconvinced that safety has been achieved—or even substantially advanced—by new designs."

Utility companies have not ordered a new nuclear power plant in ten years, and there are none in the planning stages. According to Renew America, "For utility executives, nuclear power is no longer an economical nor a prudent investment."

Finding alternative sources to fossil fuels is only part of the future of energy. Increasing energy efficiencies is another prime concern that relates specifically to the automobile industry.

By the end of the eighties there were twice as many cars—135 million—on American roads as there had been at the beginning of the sixties. Almost a third of all carbon dioxide emissions in the United States came from the burning of gasoline and transportation fuels. As a nation we drive more miles than all the rest of the world combined.

Obviously, making cars more fuel efficient and less polluting has to be a national priority. This is not to say that nothing has been done thus far. On the contrary, the gas lines of the early seventies forced an efficiency consciousness on the consumer. Where once the most important consideration in buying a car was the amount of chrome and the length of the tailfins, suddenly people began asking for better mileage. When they started buying Japanese cars to get efficiency, Detroit finally responded. By the end of the last decade the average new American car got twenty-eight miles per gallon as opposed to fourteen in 1974.

However, an unsettling trend has begun—the move to the minivan. In 1976 the light vehicle—pickups, minivans, and four-wheel-drive sports vehicles—only accounted for 20 percent of total sales. By 1989 that market share had risen to one third. These vehicles are not as fuel efficient as regular cars. In fact, they get an average 26 percent fewer miles per gallon. This has led to an increase in gas consumption.

Even if the American public does not seem as concerned with fuel economy as it once was, research in improved technologies has not stopped. The most promising ideas for lowering gas consumption in the future would include advanced engine and transmission designs, weight reduction, better aerodynamics, and energy storage innovations.

In the area of engine design, Toyota is working on an ultra-lean-burn gasoline engine that uses sensors and refined electronics. It could improve fuel usage by 20 percent. New diesel engines are being designed with ceramics that would recover energy from exhaust gases. They could improve economy by 30 to 60 percent.

The Subaru Justy already has the transmission of the future—the continuously variable transmission. In essence, it is a transmission with unlimited gears. It keeps the car's engine running at the optimal level at all times.

New materials, such as ceramics and composite plastics, will increasingly be substituted for heavier metals without a loss in safety. Auto body panels made out of these plastics have been shown to be just as strong as steel while weighing a third less.

The more aerodynamic a car, the less drag it has and the better the fuel economy. Ford has constructed an experimental vehicle that has a better drag coefficient than an F-15 jet.

Another major breakthrough to be used in cars will be the flywheel storage device. This will turn off the engine upon deceleration or braking and then use stored power to run accessories and restart the car.

Presently there are prototype vehicles using several or all of these innovations that can get up to 60 miles per gallon in the city and 75 miles per gallon on the highway. Volvo has one working prototype, the LCP 2000, that currently can get 63 in the city and 81 on the highway while being able to go from 0 to 60 miles per hour faster than the average American car. Moreover, the LCP 2000 is not dependent upon gasoline. It can use alternative fuels, such as ethanol. The car eventually will be equipped with a continuously variable transmission. Volvo estimates that they will further boost the fuel economy to more than 80 miles per gallon in the city and 100 miles per gallon on the open road.

This hatchback that seats five adults is fully developed, so why has Volvo not marketed it? "Nobody wants to buy it right now," a Volvo spokesman has explained. "You don't see a hell of a lot of commercials for high fuel economy anymore. That's not what people want to hear."

That explanation, of course, goes back to the relative cost of gasoline at the pump. That national indifference to fuel economy will change in the future as the cost to the environment continues to rise.

Future cars will eventually be weaned from gasoline, and other fuels will be substituted.

For the South Coast Air Quality Management District, that day cannot come too soon. The district, which includes the Los Angeles Basin, is attempting to improve air quality in the area. In a twenty-year plan released in 1989 it called for all new cars to be powered by something other than gasoline within the next twenty years, technology permitting. In other words, new cars sold in Los Angeles will not be allowed to burn gasoline.

Ethanol, made from corn or other crops such as sugarcane, is already being used. In 1989, however, the Brazilians experienced one of ethanol's drawbacks. The country had 4.5 million alcohol-driven vehicles on the road, but suddenly had a shortage of fuel. Sugar growers had found it more profitable to sell their crops for sugar rather than fuel.

In all likelihood ethanol production will eventually be in competition with food production. If corn fetches a higher price at the supermarket than it does at the service station, farmers will sell it as a food. The supply of ethanol, therefore, would be unreliable.

It is hoped that alcohol fuel made from wood or grasses will be on the market by the turn of the century and then will take over a large part of the market share. Researchers are now trying to create genetically engineered enzymes that will raise the ethanol yield of wood. It is believed, judging on the progress already made, that at some point there will be a 90 percent conversion rate of wood to ethanol at a cost of 60 cents a gallon—about what gasoline costs wholesale now.

Another direction that alternative transportation will take in the next few years is toward electric cars. Actually there is nothing new in the concept of a battery-powered car. Electric engines were first developed in 1916, but they never caught on because of their restricted range, speed, and durability. Now, with California leading the call for cleaner cars, interest has been renewed in electric vehicles.

The utilities industry's Electric Power Research Institute, has developed a van that will reach a speed of 72 miles an hour and travel for one hundred miles without needing to be recharged. The vehicle would be suitable for deliveries, service calls, or other short-distance functions.

Even though electric vehicles get their power from a utility that in all likelihood uses fossil fuels, the power plants are cleaner than gasoline emissions. The Research Institute, in comparing a regular General Motors van with one outfitted with batteries, found the latter to be 97 percent less polluting.

General Motors is planning to market an electric sports car called the Impact, which has been shown to accelerate from 0 to 60 faster than the Mazda's Miata and the Nissan 300 ZX. "This is no golf cart," said Robert B. Smith, GM's chairman.

However, Smith also acknowledges that at this point in development, the Impact would be expensive to maintain. At 1990 prices, in the Los Angeles area, it costs $40 a month to fuel and service a conventional car driven ten thousand miles a year. An electric car would cost twice as much. Although the cost of recharging the batteries would only average between $5 and $12 a month, the car's batteries would have to be replaced every two years at a cost of $1,500. The goal is to develop longer lasting batteries in order to bring operating costs down to a level comparable with conventional cars.

Electric cars can be recharged by running an extension cord into regular household current. Should electric cars catch on, industry officials envision "more advanced technology, like power lines buried in the pavement from which cars pick up electricity," reported the *New York Times*.

With improved battery technology, it is believed there will be a strong second-car market for electric vehicles—for around-town shopping and short-run commuting. The public has already demonstrated intense curiosity in this type of vehicle. According to Martin F. Gitten, when his company, New York's Con Edison, tested the electric car, "everywhere we took them, it was like the Pied Piper of Hamelin, it drew a crowd. There is a public fascination. All we need now is the technology."

Doug Cobb has been the object of similar fascination when he toodles around Florida in his electric car. "I get senior citizens coming up to me all the time and asking me where they can get a car like this. They think it would be perfect for them, and it would be because they're on a limited budget, and they really only need a small car to get around in."

But Cobb's electric car is different from GM's Impact or the Power Research Institute's van. Cobb, a full-time electrician and a part-time puttering inventor, developed a solar-powered electric car.

Cobb's niche in the automobile hall of fame arose out of necessity. He wanted a vehicle for his electrical contracting business that would be used for short pickup and delivery runs. Being a solar-power advocate, he decided to rig up his own vehicle. First he bought an old electric car, with twenty miles on its odometer, from the Postal Service for $800. The car had been part of a 2,000 battery-powered fleet the Postal Service had experimented with in the late 1970s. This experiment had been a dismal failure. The intention was to charge the cars overnight and use them for deliveries during the day. Unfortunately, the constant stopping and starting on a carrier's route quickly drained the batteries. Ultimately the Postal Service returned the cars to their Florida manufacturer.

Cobb acquired his car and outfitted it with a panel of photovoltaic cells on the roof. He can go between thirty and sixty miles—depending on the number of stops—before stopping to let the batteries recharge.

Oren Zimmerman of the Power Research Institute predicts that "electric vehicles are definitely going to be a reality in the 1990s. Mass-produced, solar-powered vehicles may be a little off in the distance, but there's probably a niche market for that kind of thing."

While improving fuel efficiency in cars is of major importance, the same will be done in homes and factories. Heating, cooling, and running appliances and water heaters in residential and commercial structures accounts for more than 35 percent of the energy used in the United States. So it is a rather sobering statistic that 4 percent of all energy consumed gets leaked out our windows.

We as a nation became more conservation conscious in the last few decades—weatherstripping windows, insulating attics, using the available technology and materials with some success. These technologies and materials will be vastly improved in coming years. For example, technology is now being developed that would dramatically raise the R-value—the measure of heat-flow resistance—in windows. In the early eighties something called low-emissivity (Low-E) windows were introduced that had a R-value of 3. These windows can cut heat loss by as much as 33 percent. By the end of the eighties at least 10 percent of all new homes were being built with Low-E windows.

Now superwindows with much higher R-values are being developed. There are already some with R-values as high as 12.

Also under development are vacuum insulation methods, better solar water heating systems, and new materials that would capture the sun's heat during the day and release it at night. Windows are being designed that can adjust the amount of sunlight coming into a room to reduce the amount of artificial light needed. These "daylighting" techniques will be further expanded into rooms that do not have windows. At some point, it will be possible to direct sunlight through a building to inside, core rooms.

In our factories, the thrust is to improve drivepower—that is, make electric motors and their drive systems more efficient—with adjustable speed drives and improved gears, belts, bearings, and lubricants. In addition, there is a movement toward cogeneration— the production of heat and electricity from a single fuel source. More fuel is needed when electricity is generated at a utility facility and industrial heat at the factory.

Rapid expansion of cogeneration stems from the Public Utility Regulatory Policies Act (PURPA), passed by Congress in 1978 and upheld by the Supreme Court in 1983. PURPA requires public utilities to purchase excess power from business, citizen, or other independent groups at competitive rates, whether or not the utilities need the power to meet current load requirements.

While that might seem a potential burden on the power companies, it has proved just the opposite. As Bertram Schwartz of Con Ed explained, "Some utilities need new capacity, and cogenerators relieved them of the obligation to build new power plants."

Cogeneration suddenly became more desirable to business. Any excess power generated became a potential new source of income. Large, energy-intensive industries, such as metal processors, chemical manufacturers, oil refiners, and paper processors, are all good candidates for cogenerated power. Dow-Corning of Midland, Michigan, found it was much cheaper to get electricity through cogeneration than from its supplying utility, Consumer's Power Company. Dow-Corning found they could produce their silicon for 25 percent less if they used electricity from their own 37,000 kilowatt cogeneration system rather than from Consumer's Power.

Dow Chemical Company, based in Freeport, Texas, installed a 1,300-megawatt plant that supplies all the electricity it needs with enough left over to sell to Houston Lighting and Power.

It is expected that this rapidly expanding source of relatively cheap energy will supply up to 15 percent of the nation's electricity by the twenty-first century, as compared to only 3 percent in 1981.

Eventually, electric distribution might receive the same common-carrier status that the telephone industry enjoys. This would allow cogenerators to interconnect with utility grids, saving billions of dollars in new power plant construction for utility companies. A Department of Energy study found that more than three thousand existing facilities have a capability of generating over 42,000 megawatts of electricity by the end of this century. That output would be the equivalent of forty nuclear power plants.

The preceding has been primarily a discussion of using and refining existing fuel sources and technologies. But is there a radical departure in energy production ready to explode from some laboratory? Are all the headlines about Superconducting Super Colliders and quarks and W and Z particles precursors to some new energy source? At this point, any predictions would be wild speculation.

Work with particle accelerators is all pure science, an attempt to discover more about the very nature of energy. Of course, practical applications may spring from that research, but as of now, there is no way of knowing.

On the other hand, applications from work in superconductivity are more readily apparent. Superconductors are ceramic materials that conduct electricity without resistance. When storing and using electricity, resistance creates unwanted heat and dissipates power. In theory, therefore, superconductors would allow energy to be stored and sent without diminishing its strength and at a lower cost. This could lead to cheaper electricity for home use, faster computers, and more powerful magnets, electric motors, and batteries.

In 1989 AT&T, IBM, MIT, and the Lincoln Laboratories, a lab at MIT that is government-sponsored, announced a research consortium that would look for ways to commercialize superconductivity. The consortium is hoping to receive $6 million from the Pentagon's Defense Advanced Research Projects Agency (DARPA) to add to the $10 million the four consortium members will be contributing.

EIGHT

Medicine on the Threshold of a Golden Age

GERONTE: It seems to me you are locating them wrong: the heart is on the left and the liver is on the right.

SGANARELLE: Yes, in the old days that was so, but we have changed all that, and we now practice medicine by a completely new method.
—MOLIÉRE, *The Doctor in Spite of Himself*

MEDICINE and its practice have been modified, revolutionized, evolutionized, and turned upside down innumerable times in the course of man's history. It has been a discipline filled with trials and many errors.

Trepanning, for example, must have seemed a reasonable and correct procedure to the American Indians trying to alleviate the excruciating pain of a headache sufferer. Open a hole in the poor person's skull and let the evil spirits out. The procedure was actually of some benefit to the victim-patient who happened to have a brain tumor. Of course, it killed most of the others.

The ancient Egyptians were among the first to use drug therapy.

Trial and error found that some herbs eased pain and treated symptoms. Even today 40 percent of all drugs on the market come from naturally occurring substances. (Before we give the ancient Egyptians too much credit, it should be remembered they also used a mallet as an anesthetic.)

Under the Code of Hammurabi, the Babylonians had their own version of malpractice insurance. It was literally an eye for an eye. If a surgeon made a mistake resulting in the loss of a patient's eye, the surgeon lost his eye as well.

It took the Greek Hippocrates to usher in the first stage of modern medicine. He did so by finally making a distinction between medicine and superstition and by setting forth a code of ethics for doctors. While the Greeks naively attributed disease to the four "humors" of blood, black bile, yellow bile, and phlegm, they greatly contributed to the understanding of the body by dissecting cadavers for teaching purposes. The Greeks had the heart on the left and the liver where it should be.

And so it went through the centuries, advances and setbacks, setbacks and advances. While Renaissance doctors were bloodletting, Ambroise Pare was using a sophisticated method for stopping hemorrhages.

Discovery built upon discovery—Harvey and the circulation of blood, van Leeuwenhoek and the microscope, Stephen Hale's demonstration of blood pressure—all leading to the first golden age of medicine that began in the nineteenth century and ran into the twentieth. The role of bacteria in disease was discovered. New vaccines were found, and diseases such as typhoid fever, whooping cough, polio, cholera, bubonic plague, and measles were at last brought under some degree of control.

Surgery was no longer a butcher's practice, but a technician's art. Organs were transplanted, cancers removed. Diagnostics became highly refined as doctors were able to peer into the patient's body using computerized axial tomographic (CAT) scanning, nuclear magnetic resonance (NMR), positron emission tomography (PET), and magnetic resonance imaging (MRI).

Yet, despite all the advances, all the high technology, all the libraries of knowledge, medicine is still a science of treatment. But that is about to change.

A veteran newspaper reporter specializing in the medical field recently confessed to a younger colleague, "I envy you. You will be writing about cures."

We are entering the age of genetic medicine, a period so full of promise and rapid developments that it has been compared to the time one hundred years ago when every medical journal was filled with reports of new bacteria that had been linked with new diseases.

There are 3,500 known genetic diseases:

- 2.5 million people suffer from Alzheimer's
- 2 million from manic depression
- 100,000 from neurofibromatosis, or Elephant Man's disease
- 50,000 from Duchenne muscular dystrophy
- 27,000 people are stricken each year with melanoma

This is only a partial list. It is estimated that twelve million people have some kind of genetic disorder and tens of millions more have illnesses that are affected by genetic factors, such as cancers of the lung, breast, colon, rectum, and cervix.

The ultimate goal of genetic medicine is to terminate diseases at their genetic roots. Instead of using partial treatments, such as surgically removing cancers or using drugs to arrest development of illness or alleviate its symptoms, doctors will be able to drive out the malady completely.

Before this can happen, the locations of the defective genes have to be found, which is not an easy undertaking.

Take one example: In 1987 Doctors Francis Collins of the University of Michigan and Dr. L. C. Tsui of the Toronto Hospital for Sick Children set about searching for the gene that causes cystic fibrosis, a disease that affects the lungs and usually causes death before the patient is thirty years old. CF is the most common birth defect in the United States.

That it only took eighteen months to track down the gene seems next to miraculous when you consider that there are between 100,000 and 300,000 human genes arranged along twenty-three pairs of chromosomes. By the end of the last decade, only 170 of the genes responsible for the thousands of hereditary diseases had been pinpointed.

In the future, doctors working on particular diseases will not have to go on the tortuous journey that Collins and Tsui followed. Instead, they will have at their disposal genetic maps thanks to the Human Genome Project, a $3 billion federally funded endeavor directed by James Watson, who was co-winner of a Nobel Prize for the discovery of the double-helix structure of DNA. Tens of thousands of scientists will take part in the project.

Watson is intent on completing a detailed map of where all the 100,000 or more genes are located on all forty-six chromosomes within five years. Furthermore, he believes that within another ten years the sequence—or precise order—of all the three billion nucleotides (the chemical units that make up the genes) will have been determined.

There are many ways all this genetic information will eventually be used. Scientists will be able to develop prenatal tests to determine whether a fetus is suffering from any genetic disorders.

Diagnostic tests will be created that will determine whether someone has a particular disease or is likely to get it later in life.

It is expected that thousands of "protein" drugs will be developed. The body manufactures small quantities of certain proteins such as interleukin-2 and interferon, which fight cancer. Once scientists have found the gene that controls the productions of such proteins, that gene could be placed into bacteria and the protein produced in greater quantities.

Perhaps the most exciting outcome of gene mapping and sequencing, however, will be gene therapy. With this, defective genes will be repaired or replaced, thus returning the patient to good health.

"When we have sequenced the genome, we will have developed a powerful set of tools that will prepare us for the twenty-first century of biology," said Charles de Lisis, a former Department of Energy administrator who was active in getting the Genome Project off the ground.

Another historic step in genetic medicine was taken at 10:47 A.M. on May 22, 1989. It was precisely then that a nurse in the surgical intensive care unit at the National Institutes of Health switched an intravenous bag for a patient. The first gene transfer experiment on a human being was launched. Genetically altered cells were being injected into a patient with skin cancer.

W. French Anderson, chief of the laboratory of molecular hematology at the National Heart, Lung and Blood Institute and one of the leaders of the experiment, described the experiment as being "like a moonshot. Everything went exactly according to plan because we had gone up to the point of putting the cells in a patient at least a dozen times before."

What had been put into the patient was actually a tracker gene, tailored to follow the movements of a certain kind of white blood cell, called a T cell, as it coursed through the patient's body in search of tumors to destroy. Besides adding valuable information to a cancer study, the experiment was conducted to see if foreign genes could be transferred successfully. If the patient had died, Anderson explained emphatically, "that would have set the whole field back."

But the patient did not die, and more gene transfer experiments—leading us closer and closer to gene therapy—are scheduled. Anderson predicts that the first real gene therapy will be aimed at curing cancers of the eye, breast, bone, and lung by replacing a missing gene. These cancers strike when the retinoblastoma gene is lost from the cell. It is believed that this gene can prevent the uncontrolled growth characteristic in cancer. Experiments with animals have demonstrated that inserting the retinoblastoma gene into cancer cells can stop the formation of tumors.

As T. Friedmann, a gene therapy researcher at the University of California at San Diego, has commented, "We are arguing that you fix what is broken. Cancer is a genetic disease. If the genes are broken, you fix the gene."

Some scientists are suggesting sequencing the genome might be used to find answers to basic questions of science, one of which is what causes an egg to grow into a human. "If you could understand that in detail," said Nobel laureate Walter Gilbert, "you could regrow any part of the human body."

As Victor McKusick, medical genetics professor at Johns Hopkins University, has said, "Today's science fiction tends to be tomorrow's science."

Moral and ethical questions, however, have been raised about gene therapy. Some people fear, for example, that prenatal testing could lead to unwarranted abortions. Marc Lappe, director of the Humanistic Studies Center at the University of Illinois in Chicago, has wondered if parents, upon finding out that their unborn child would be afflicted with bipolar disorder (characterized by periods of depression and mania), might decide to terminate the pregnancy. If such tests had been available in the past, Winston Churchill, who was afflicted with bipolar disorder, might never have been born.

"Pressures to use information from prenatal diagnoses will be great," Gilbert has conceded. "Our society is already confused about whether to have an abortion if the child is likely to have Down's syndrome. What will happen when they know hundreds of details about their newborn? Society will have to come to grips with the abortion issue."

There seem to be positives and negatives to almost all aspects of genetic medicine. For example, it is predicted that in the next century everyone at some time in his or her life will be given a genetic profile. From the profile, the individual will be able to ascertain what tendencies he or she might have toward certain diseases. Someone at high risk of getting skin cancer, for instance, would be advised not to get a job as a lifeguard.

What if employers or insurance companies use genetic profile information against someone, to deny them a promotion or coverage?

Watson, recognizing these concerns, has already earmarked between 3 and 10 percent of his Genome budget to address the moral, ethical, and legal issues. He has appointed Dr. Nancy Wexler, an expert on Huntington's disease at Columbia University, to chair an ethics committee.

Among the questions that Wexler is asking are:

- If genetic tests are available, should they be voluntary or not?
- Can insurance companies require applicants to submit genetic tests to get coverage? Can they charge exorbitant premiums for high-risk individuals?
- If insurers reimburse parents for a prenatal test, have they the right to require termination of an affected fetus? Can they refuse to pay health-care costs for mother and child?
- Can employers require testing as a condition of employment? Should someone with a genetic predisposition to a heart attack be allowed to be a pilot?

While it may take years before we enjoy the most important fruits of the Human Genome Project, something called monoclonal antibodies are being used today.

These are laboratory-designed antibodies that selectively bind to disease-related cells, unlike naturally formed antibodies which attack all foreign intruders within the body.

First synthesized in Cambridge, England, monoclonals have been found to be effective treatment for liver cancer, where remission time in one study was tripled and the size of the cancers greatly reduced. Patients are given injections of monoclonals that contain radioactive iodine 131. The antibodies then seek out the cancer cells and chemically bind themselves to the cancerous growth. Then the irradiation process begins.

Some liver patients treated with monoclonal antibodies have lived cancer-free for almost four years, experiencing tumor reduction of fifteen pounds.

Doctor Stanley E. Order, professor of Radiation Oncology at Johns Hopkins, has said, "We have the first effective treatment for liver cancer . . . monoclonal antibodies are the wave of the future."

In 1988, scientists at the Molecular Biology Laboratory in Cambridge also reported success with leukemia. Two patients recovered enough to leave the hospital after receiving the "magic bullet" (monoclonal) therapy.

Dr. Mike Clarke, part of the Cambridge team, would not go so far as to say there had been a cure, but he did report that "all the cancer cells we could see have been destroyed. The significance of this is very great."

The monoclonal field is still relatively new, but already two competing drugs have been developed to treat a syndrome called septic shock. Until recently, this was a malady that most people had never heard of, although its most common type kills eighty thousand Americans a year, or more people than have died in the U.S. from AIDS since that epidemic began. Symptoms of septic shock include fever, confusion, and chills. Often doctors do not make a proper diagnosis at the onset of the problem. By the time they do, the patient has experienced kidney, lung, and liver failure, and it is too late to save him.

The deaths of two widely-known people brought septic shock into the national spotlight. Heather O'Rourke, the young actress who had intoned "They're heeeere" in the movie *Poltergeist,* was struck down by this disease. Septic shock also contributed to the death of former Studio 54 owner Steve Rubell.

Before the arrival of the two monoclonals, most of the drugs used for septic shock were antibiotics and were only partially effective. Tests have shown that using one of the monoclonals, Xomen-E5, reduces the mortality rate by 40 percent if the patients have not already gone into deep shock.

Dr. Charles J. Fisher, Jr., chief of critical-care medicine at Case Western Reserve University School of Medicine in Cleveland, has said, "If these agents [the new monoclonals] work as one hopes they do, they are going to have the impact like when penicillin was discovered and started to be used. That is not loose talk."

Early projections were that there could be a $500 million market for these drugs in the United States and more than a $1 billion market worldwide.

The one marked difference between Xomen-35 and its counterpart Centoxin is that the latter is cloned from a human cell and the former from that of a mouse. Monoclonals derived from mouse cells are sometimes rejected.

Rather than resorting to human cells, some scientists have turned to plants. According to Andrew Hiatt, a molecular biologist at the Scripps Research Clinic and Foundation in La Jolla, California, where alternative technology is being refined, cloning from plants could lower the cost of monoclonals from $5,000 to 10 cents a gram, as well as reduce the risk of rejection.

With the concern over rejection, it may sound somewhat surprising that monoclonals are being suggested as a new method for stopping the rejection of transplanted organs. But scientists at Stanford University have had some success in using specially designed monoclonal antibodies to temporarily "blind" rats' immune systems and keep them from recognizing transplants as foreign objects.

Besides their use as a drug, monoclonals are used in the diagnosis of a variety of diseases, including AIDS. They are also used in home pregnancy tests.

Before looking at specific illnesses and what kind of progress we can expect in their treatment, there are two other fields that deserve at least a few words.

One is nanotechnology (which was already mentioned in the discussion of computers). Admittedly, this field is only in its earliest stages, but there are high hopes and predictions about its eventual impact.

Should nanotechnology reach its potential, we can expect to see a wide range of applications in medicine. Nanotechnology advocate K. Eric Drexler suggests that computers could be built "small enough to fit in one-millionth of the volume of the human cell, and then direct molecular machines to sense structures inside the cell and repair them."

Just as nanotechnology could manufacture a tomato from the

molecule up, so could it make human replacement parts, such as hearts and toes.

Before nanotechnology reaches that degree of advancement, the interim technology will be that of cyborgs. Artificial parts with skin-like coverings will be run with such advanced and sophisticated computers that the interface between body and machine will be virtually invisible. It may be impossible to tell a cyborg part from the real thing.

Such cyborg technology will not be limited to replacing damaged organs or limbs. As a person grows older, the human body will give way to cyborg parts until the one remaining original part of the person would be the brain. According to Hans Morevac's conjectures, the mind, too, might eventually be transferred into a machine. Of course, the chances of any of this coming to pass in our lifetimes or our children's children's lifetimes are remote.

Let us turn to items closer to us—diseases and methods for diagnosing and treating them that will be emerging from our laboratories.

AIDS (Acquired Immune Deficiency Syndrome)   AIDS is a truly devastating worldwide epidemic. In the United States alone there have to date been over 120,000 cases of AIDS reported, 70,000 of those already having resulted in death.

While each day brings new developments and theories, and anything written now may swiftly be out-of-date, it is worth mentioning a few promising drugs. An experimental vaccine, HPG-30, will undergo extensive human testing in Europe throughout the next few years. At the National Institute of Allergy and Infection, testing of another experimental vaccine, GP160, has already produced an immune reaction in volunteers. In addition to the widely-used AZT, drugs such as GLQ223, Compound Q and alpha interferon are now subject to intensive testing. Genetech's synthetic drug CD4, an exact copy of proteins on the walls of the immune cells, may allow other drugs to be delivered straight to the AIDS virus. As for immunization research, Dr. Jonas Salk and his team are experimenting with a program that involves injecting uninfected subjects with the AIDS virus stripped of its coat.

But, despite the great effort and many millions of dollars already dedicated to AIDS research, much more needs to be spent on care, prevention, treatment and the discovery of a cure.

Alzheimer's Disease   An estimated 2 million Americans, mostly elderly, are afflicted with this cruel brain disease which often destroys its victims' memories and their ability to care for themselves.

In 1989, researchers were able to find a protein linked to the disease in a site outside the brain. The substance, amyloid beta protein, was found in blood vessels, under the skin, and in the colon of Alzheimer's patients.

Scientists say this finding could lead to a laboratory test that would confirm diagnosis of the disease. It might also aid in the development of treatment.

Arthritis (Rheumatoid)   This is a painful condition caused by the body's rejection of its own healthy tissue in the joints. The standard treatment for the more than two million arthritis sufferers in the United States has been the use of painkillers like aspirin or ibuprofen or anti-inflammatories like steroids.

However, researchers now report a protein that checks inflammation and actually halts decay of the cartilage. This substance, discovered by scientists at Synergen, a Colorado biotechnology firm, works by curtailing the action of interleukin-1. Under normal conditions the hormone interleukin-1 fights infections. But in arthritis sufferers, the hormone goes out of control.

According to Dr. Robert C. Thompson, who was in charge of the research team, the protein is "not going to repair the damage that's already been done, but it's going to stop the disease from getting worse."

Meanwhile, in Berkeley, California, the Xoma company is conducting clinical trials on a drug that combines a monoclonal antibody with a toxin; Xoma hopes the drug will be effective in treating rheumatoid arthritis.

Back Pain   Chymopapain, a drug that was withdrawn from the market for five years because of adverse side effects, is now receiving renewed interest as an alternative to surgery for people with spinal disk problems.

The drug, which is derived from papayas, had been used for treating ruptured spinal disks. When some patients experienced nerve damage and allergic reactions, many doctors stopped prescribing it. Subsequent studies discovered that many of the problems stemmed from faulty injections.

Bacterial Infections Monobactums, a new class of bacteria-fighting antibodies, are now being used in the treatment of a variety of ailments including certain forms of pneumonia, bronchitis, peritonitis, endometriosis, and cystitis. One of the first monobactums is called Azactam, marketed by the Squibb Corporation.

Blood Substitutes The tainting of the nation's blood supply with the AIDS virus, as well as shortages of donors, has given a new push towards finding an acceptable blood substitute for certain types of transfusions.

Biotechnology firms are experimenting with cows' blood, or at the least with the part of it that carries oxygen—the hemoglobin. Usually the human body would reject a foreign substance, but there are exceptions, such as cow insulin, which diabetics tolerate. Scientists hope the cows' hemoglobin will be another exception.

Among the advantages to using bovine hemoglobin is its long shelf life. Human blood must be thrown away after three weeks. Cows' blood can be kept for six months.

If the bovine hemoglobin tests well, the military would like to freeze-dry it for use in combat zones where it could be reconstituted with sterile water. Paramedics could use it in ambulances rather than having to wait until they reach the hospital.

Cancer As previously mentioned, the great hope for the future of cancer patients lies in gene therapy, but in the meantime, monoclonal antibodies are being used and others are being developed to treat specific cancers.

Also on the horizon is a potential vaccine against cancer. Frank J. Rauscher, Jr., senior vice president for research for the American

Cancer Society, said in 1988, "If we compare where we are now in the war on cancer in the 1980s with where we were in the war on polio in the epidemic summers of the 1950s, we are perhaps five years away from the final study that brought us a safe, effective vaccine. The analogy is not perfect. Cancer, after all, is not one disease but more than one hundred. But there are characteristics common to all cancers. The problem is not one hundred times more complicated. We are entering a Golden Age of discovery."

The lives of three thousand colon cancer victims may be saved each year because of a new form of chemotherapy. Some 53,000 Americans die yearly from this form of cancer, making it second only to lung cancer. Researchers believe the administration of two drugs, levamisole and 5-flurouracil, after surgery will reduce the death rate by 10 to 15 percent.

Results from the studies on this treatment prompted the National Cancer Institute to mail a special announcement in 1990 to 36,000 physicians and researchers urging them to use the treatment as soon as possible.

Preliminary studies have suggested that cruciferous vegetables— broccoli, kale, and cauliflower—might help prevent cancers in general. Other studies indicate that a natural, nontoxic form of Vitamin A, beta-carotene, may heal mouth lesions that can develop into neck and head cancers.

Heart Disease In 1985 alone, 1.25 million Americans suffered a heart attack. Some 550,000 of these victims died, 350,000 without even reaching the hospital.

Among new experimental procedures is the use of lasers to repair small blood vessels. A second technique uses lasers to vaporize plaque on vessel walls, thus creating enough space for blood to flow adequately.

Hiccups Uncontrollable bouts of severe hiccups are no laughing matter. Studies are indicating that the heart drug nifedipine may provide relief for some chronic hiccupers.

Parkinson's Disease Parkinson's, a brain disorder affecting 400,000 Americans, most often those over fifty years old, is a slow, usually fatal disease that causes muscle tremors, stiffness, and weakness. In 1989, it was discovered that the drug selegiline, which is sold under the names Deprenyl and Eldepryl, will slow the progression of the disease.

In a study in which fifty-four patients participated, selegiline usage doubled the time between the onset of Parkinson's and the point the victims had to turn to therapy using the drug L-dopa. While L-dopa does initially alleviate some of the disease's symptoms, it loses its effectiveness after only a year. It can also cause side effects such as abnormal movement and hallucinations.

Skin Healing   An experimental hormone called epidermal growth factor has been found effective in making skin grow more quickly. Because it is produced naturally in only small amounts, the hormone has been produced in quantity through genetic engineering. It may eventually be used to speed healing of wounds ranging from minor cuts to life-threatening burns.

Strokes   A new class of drugs called lazaroids may be able to stop brain cells from dying after a stroke. The drugs also hold potential for treating Parkinson's, Alzheimer's, and other brain disorders.

Tooth Decay   There may come a day when customized viruses are in the vanguard of fighting tooth decay. Alan H. Norris, a chemical engineer, has patented just such a method.

Weight Control   First the bad news: There is still no magic pill that lets you eat *anything you want* without putting on an ounce.

But there is Olestra, or at least there will be if the FDA ever manages to wade through eleven thousand pages of data Proctor and Gamble submitted with its petition for approval.

P&G explains that "Olestra looks like fat. It tastes like fat. And it cooks like fat. In practical terms, when used in cooking and frying, it is the same as fat, because it is made from fat. But it differs from ordinary fats or oils in one respect. Instead of a molecule of gylcerin in its middle, olestra has a molecule of sucrose—table sugar. This makes a more bulky molecule which the body cannot break down, cannot absorb into the bloodstream and cannot convert into calories."

While you are waiting for Olestra to hit the market, you might want to pick up some Fat Magnets, which, because they are made from only natural ingredients, do not need FDA approval. These pills are supposed to keep the body from absorbing some of the fat and cholesterol in foods. The developers, two cardiologists from Los Angeles, claim that patients who took the Fat Magnets while on a 1,200-calories-a-day diet lost on an average of 8.1 pounds over a three week period, while patients on the same diet who did not take the magnets only lost 3.8 pounds.

NINE

Space: Getting Acquainted with the Unknown

We are a nation of explorers. It's a fundamental thing to want to go, to touch, to see, to smell, to learn, and that I think will continue in the future.

—NEIL A. ARMSTRONG, *on the twentieth anniversary of his first steps on the moon*

FOR MOST of man's brief existence, he has been bound to Earth. At first the shackles were forged by his own lack of vision and ability. As science progressed, theories were expounded. Pythagoras, in the sixth century B.C., suggested a spherical earth and a universe filled with objects bound by the laws of nature. In the second century A.D., Ptolemy got things turned around a bit and put the earth at the center of the universe.

That is where we stayed for fourteen centuries, until the Polish astronomer Nicolaus Copernicus had the temerity to suggest the earth and its satellite moon actually circled the sun. His heliocentric theory brought on the modern age of astronomy and ushered in the incredible seventeenth century. It was then that Johannes Kepler discovered the theory of planetary motion, Galileo began using the

143

telescope to observe the heavens, and Sir Isaac Newton formulated the laws of gravity.

Yet for all the information and knowledge that continued to build on this base, we remained locked to the Earth until the twentieth century. It was then that such scientists as Konstantin Tsiolkovsky, Robert Goddard, and Hermann Oberth began toying with the idea of developing tremendously powerful rockets that would thrust a spacecraft beyond the grip of gravity. It was the Germans who threw their research and deutsche marks behind rocketry, leading to the development of the V-2 in World War II.

Postwar years witnessed the all-out race between the Soviets and the U.S. to be the first in space. Rocket after rocket was fired straight up in the sky to see how far and fast it could go.

And then up went Sputnik, the first artificial satellite to orbit the Earth.

The race heated up. Explorer. Jupiter. Vostok. Mercury. Yury Gagarin. Alan Shepard. John Glenn. Valentina Tereshkova. Gemini. Voskhod. Soyuz. Apollo.

Finally, at 2:56 A.M. Greenwich mean time, on July 20, 1969, a man stepped onto the surface of the moon.

We had gone a long way, literally and figuratively, from Ptolemy and Copernicus. Neil Armstrong had taken that "one small step for man, one giant leap for mankind," and then we seemed to lose our way. Maybe it was because we had such faith in NASA as the one entity in government we could rely on, that knew what it was doing and got it done properly, that the space program became almost boring. There was no longer any urgency. No one, except presumably those at NASA, knew quite what was on the agency's agenda. It lacked a grand plan such as "Beat the Russians to the Moon" or "Get out there and explore the final frontier; go where no man has ever gone before."

Besides, there were all those distractions that took up the nation's energies and attentions: Vietnam, Watergate, recession, gasoline lines, Iranian hostages. Even the launch of the space shuttle on April 12, 1981, failed to capture our imagination and give us a sense of purpose in regards to the exploration of space.

The space program, when compared to the hyperspace swash-buckling of Luke Skywalker in *Star Wars*, had become boring. Watching the space shuttle launched from Cape Canaveral once or twice was enough. And the never-ending stories of glitches and heat tiles needing replacement grew tedious.

Flight 51-L was to end our apathy. Ninety seconds into the launch of the space shuttle Challenger, we as a nation became mesmerized once again by the space program. But for all the national agony the explosion may have caused, an event that will have a longer-lasting and more indelible effect on the country's future actually occurred the previous year.

For it was in 1985 that Ronald Reagan appointed a National Commission on Space to formulate a "bold agenda" for a civilian space enterprise. After a year of studies, hearings, proposals, and all else that goes into such a report, the commission issued a goal for America "to lead the exploration and development of the space frontier, advancing science, technology, and enterprise, and building institutions and systems that make accessible vast new resources and support human settlements beyond Earth's orbit, from the highlands of the Moon to the plains of Mars."

In other words, the plan was for man to move permanently beyond the confines of Earth. What made this report so promising was its considered, stepped approach. No more flinging ourselves haphazardly out into space for sheer exhilaration and bragging rights. No more barrelling off for the moon or Mars for public relations purposes.

Instead, the plan for the future would be phased and gradual, pushing the boundaries, establishing bases, exploring further, while at the same time opening the way for commercialization. It might seem, at first blush, that calling for commercialization was a ploy by NASA, whose budget is dependent on the backing of the president and the whims of Congress, to get support from the business community. In fact, it was probably more an acceptance of the economic nature of the nation. Sooner or later, we demand a payback. It is all right to spend the millions on research and start-up, but eventually the endeavor must be thrown open for public profit-making.

For a while there was some question as to where President Bush stood concerning space exploration. Would he come out for the commission's proposals? Would he scrap colonization of the moon and go straight for Mars? Would the moon be far enough? Bush answered these questions in a speech celebrating the twentieth anniversary of Neil Armstrong's stepping onto the moon. He had bought the whole package.

"Our goal," he said in the summer of 1989, "is nothing less than to establish the United States as the preeminent spacefaring nation. . . . And next, for the new century, back to the moon, back to the future, and this time back to stay . . . And then, a journey into tomorrow, a journey to another planet, a manned mission to Mars."

To make this march into space work, a jumping off place is essential, and Earth is not it. Building everything here and then launching it out of the gravitational pull is an expensive proposition. It will be far more efficacious and cost-effective to manufacture and build away from Earth. The first step towards that goal is the manned space station, Freedom. Freedom will be quite unlike Skylab of the seventies, which was nothing more than a temporary base for three-man crews to do some fast experimenting. The longest occupancy of Skylab was eighty-four days. Skylab, a refitted third stage of a Saturn V booster, was little more than a lean-to in comparison with what Freedom will be.

When its first stage is finished, the space station will resemble an enormous extraterrestrial insect. Its wings will be a 508-foot horizontal truss of incredibly strong and lightweight graphite-epoxy pipes. The pipes are designed to snap together like a children's toy.

At either end of the horizontal trusses will be four solar panels that will absorb the energy of the sun with 250,000 solar cells. The eight panels combined would cover a half acre and would generate enough power for fifteen average American homes.

Freedom's body will be made up of modules for living and

working. Initially, six are planned. At the bottom will be two for NASA, about forty-five feet long. One will be living quarters for eight.

The second NASA module will serve as a laboratory. Both the lab and living habitat will have docking nodes so that crews can enter or leave shuttles.

On the top of the truss will be a forty-foot European Space Agency module. It will combine both living and lab facilities where personnel will study material processing, life sciences, and fluid dynamics.

Next to this will be three smaller, piggybacked modules. One will be a one-person Japanese live-in lab for material processing, communications equipment, and engineering studies, with the other two modules serving as storage units.

Freedom will be constructed in stages. NASA hopes that by late 1995, an unmanned cargo shuttle will blast off from Cape Canaveral, carrying the first load of construction girders, electronic equipment, and solar panels. The cargo will be released into a low orbit. In a few weeks another cargo will be blasted up, at which point astronauts will begin construction of Freedom.

It will take twenty such cargo deliveries and three years before our first permanent outpost in space is ready for occupancy. And then, Congress and its purse strings willing, man will be truly ready to step out into space.

"In the short run we have two exploration goals," Frank Martin, assistant administrator for exploration at NASA, told *Discover* magazine in 1989, "the moon and Mars. All roads to these places will start on Earth and pass through the space station."

It is NASA's plan that eventually the crew of the space stations will put together pieces of exploratory spacecraft brought up by unmanned shuttles or disposable boosters. "In essence," Martin said, "the space station would become a service station."

The Apollo astronauts touched the surface of the moon, gathered their rocks, set up machines for gathering data, hopped back into the lunar module, and made a beeline for home. They provided great photos and sound bites for the evening news and great publicity for NASA. However, except for getting the country excited, it was not the most efficient way to open up the moon to enterprise and colonization.

Over the years there has been a great deal of argument over how best to proceed. One faction questions the efficacy of manned missions in general, suggesting that space exploration, at least into the early part of the next century, should be left to machines. Then there was the question of bypassing the moon altogether and going directly to Mars. Of course, there were those who wondered if we should leave Earth at all, and instead invest the billions on solving problems closer to home.

While no definite answer has been reached, it looks as if the next missions to the moon will be unmanned. Robotic explorers will be sent to gather data. Their first priority will be to determine what lurks in those shadowed and mysterious craters near the lunar polar ice caps.

In the early 1990s, the long-delayed Galileo spacecraft destined for Jupiter will finally be launched. On its swing past the moon, a special probe will search for ice glistening in starlight.

Next on the agenda will be the Lunar Observer, scheduled for the mid- to late-nineties. It will be put in a lunar polar orbit, taking photos of possible landing sites, collecting geological data, and, again, searching for any signs of water.

After that, robotics will be landed to try to penetrate the moon's surface to gather still more information concerning the composition and treasures of our natural satellite.

Then it will be time for man to return and truly discover what the moon has to offer.

This time, unlike our earlier forays, men and women will stay for extended periods on the moon, first in outpost camps where the astronauts will have specific study tasks. These will include seismology, land, and resource surveys. These early moon pioneers will serve as scouts.

Once their data is gathered and assessed, it will be time for the construction of permanent base camps and lunar towns that will not only supply the outposts, but will have medical and research facilities as well. How man functions away from Earth for long periods, how his body and mind adapts, will be examined on the moon. Once the moon is ours, we will be ready to move on into the solar system.

Venus is one of the two closest planets to Earth. Data sent back from our Pioneer and Mariner spacecrafts and the Soviet's Venera spacecraft have shown that it is an inhospitable environment for man, a place we would not want to visit or live.

But what of the other close planet, Mars? It has long fired our imaginations. When we realized the red glow in the night sky was a planet, we looked to it as our neighbor, where other beings might live, beings whom we might contact and learn from.

For many people, the idea of communicating with Martians became an obsession. In 1820 a German mathematician wanted to let the Martians know we were here. He came up with a plan whereby a gigantic triangular field of wheat, surrounded by pine trees, would be grown in Siberia. He reasoned the Martians would recognize this signal as a sure sign of intelligent life on Earth and would respond.

A Frenchman had a different plan. He wanted to build an enormous mirror that would burn a message in the sands of Mars with reflected sunlight. Then there was the inventor who wanted "to talk to the planets." His idea was to send electromagnetic surges from his laboratory in Colorado using an electric cord, seventy feet in diameter. There is no record of his receiving any message in return. He did, however, set light bulbs glowing for some twenty-five miles around.

We turned our telescopes to Mars and saw "canals" ribboning its surface and were certain great civilizations had constructed them. We have made the planet part of our lore and legend, and now we are set to make it a waystop on our "highway to heaven," as the space commission put it.

Pictures of Mars's surface sent back in 1965 by Mariner 4 dispelled many of our romantic notions of what is actually there. The Viking landers in 1976 only confirmed those first fuzzy pictures. No remnants of glorious ancient civilizations. No canals. No vegetation. No life.

The landers gave us a view of a landscape dotted with boulders, craters as large as the United States, mountains five and a half times higher than Everest. Sand storms, racing at half the speed of sound, roar for months on end. Its temperatures drop to as low as -184 degrees Fahrenheit and as high as $+68$ (although that is only at one equatorial "oasis"). There seem to be polar caps of dry ice and perhaps even permafrost. Mars has clouds and fog, pink skies, and reddish-gray soil.

There is gravity (only one-third that of Earth's) and an atmosphere (a meager one percent of what it is at Earth's surface). Still, compared to Venus and our own moon, it is an enticing, friendly environment.

Exploration of Mars actually began in 1976 when Viking landers transmitted images from the planet's surface and sky. The Russians tried unsuccessfully to send probes to Phobos and Deimos, the moons of Mars. They had hoped to fire laser blasts in order to analyze the resulting vapors. But problems kept the mission from succeeding. Rather than attempt another similar one, the Soviets are embarking on a far more ambitious project. In collaboration with the French and an independent group called the Planetary Society, Russia will send balloons to the skies above Mars. The balloons will carry cameras capable of photographing objects as small as four inches on the planet's surface.

For ten days the balloons will be carried around the planet by the Martian winds, whizzing past volcanoes three times the size of Mount Everest, and flying through canyons far deeper than the Grand Canyon. The data collected will be used later to find suitable landing and base sites for unmanned and manned forays.

At night, when the fierce winds are silent, the balloons will hover, almost stationary, as a "snake" attached to them slithers along the surface collecting bits of dust and rock for later analysis. The snake is being designed by scientists at the California Institute of Technology. Bruce Murray, who is heading the snake project, says it will consist of a chain of titanium shells, "like a long chain of tin cans."

(An interesting aspect to this mission is that it is believed to be the first time a private organization became involved in interplanetary exploration. The Planetary Society is a group of 125,000 scientists and space enthusiasts.)

Information from the balloon runs and other robotic probes will help fulfill the National Commission on Space's prediction that by early in the twenty-first century permanent bases on Mars will be established. If man is to live on Mars, however, new ways of sustaining life will be necessary. Both the U.S. and the U.S.S.R. have been exploring one avenue—biospheres. These are, essentially, environments in bubbles. Enclosed, they are self-maintaining and support a variety of plants and animals. A project named Biosphere II (the Earth being Biosphere I) has already been built near Tucson, Arizona. It is an airtight building of glass and steel that will cover four acres.

In September, 1990, it is planned that four men and four women will enter the biosphere and live in it for two years. They will grow their own food, and their oxygen and water will be produced within the artificial ecosystem. They will not be entirely cut off from Biosphere I, however. There will be contact with outside scientists, and the biosphereians will not have to give up television or radio. If this and other enclosed environment experiments prove successful, it is expected they will dot the Martian landscape within the next two hundred years. Once a feasible way of sustaining life as we know it is perfected, true colonization can begin.

Most of what has been discussed here can be accomplished with existing technology. But these technologies were already showing their age by the end of the 1980s. Rockets were still blasting off by burning liquid and solid fuels to create energy, a very expensive operation. More advanced, cheaper methods were being sought.

The Air Force is doing research with an anti-matter motor that creates energy by smashing protons and antiprotons against each other.

The military has also been working on "railguns," which use electromagnetic catapults.

Solar rockets that would use the sun's rays to heat hydrogen are being researched. Renewed interest has been shown in nuclear motors. This type had been worked on successfully from the mid-fifties until the early seventies. Hydrogen is heated by moving it around radioactive pellets or a nuclear core.

Laser boosters are also being considered. A powerful laser, either sitting on top of a mountain or on a space platform, would be aimed at the spacecraft to heat a supply of hydrogen, perhaps in the form of a block of ice.

At this point, it is unclear which, if any, of these approaches to new power sources will prove successful. What is clear, however, is that new technology is vital. A 1987 report that came out of the Washington-based National Research Council warned that "no new rocket engine development has been initiated for at least seventeen years. A revolutionary approach to advanced propulsion concepts is essential if the United States is to regain its world leadership position in space."

Railguns, solar rockets, and the like all deal with breaking free of the earth's gravity. But we will also require new technologies for moving around more rapidly out in space. Currently it would take one year to get to Mars and six to get to Jupiter.

Some of the more promising technologies being entertained now include electromagnetic accelerators, the "ion engine" for missions to the outer planets, and the "mass driver" for moving around closer to the sun.

The ion engine would get its power from ions of cesium, argon, mercury, or perhaps even oxygen being accelerated by electrostatic fields. It does not produce a powerful thrust, but would give steady propulsion over a long period. This type of engine has been used in space, but would need adaptation for the projected far-planet missions.

The mass driver would use magnetic fields to accelerate just about any material from moon soil to pulverized shuttle tanks that have been discarded. It provides a strong thrust needed for launching payloads from the moon or asteroids.

Scientists at the Massachusetts Institute of Technology have predicted that mass drivers can reduce payload costs from the $325 per pound on the shuttle in 1986 to only $1 per pound.

In the meantime, we will continue to explore the solar system. Even if man himself does not go physically farther, he will be sending his agents.

For example, the Galileo probe (the 2.5 ton spacecraft with a price tag of $1.4 billion) should reach Jupiter in 1995.

When it gets there it will probe the dense, hydrogen atmosphere, searching for clues of the origin of the universe.

"Jupiter is most important to us in trying to understand how the solar system evolved," said William J. O'Neil, Galileo science and mission design manager at the Jet Propulsion Laboratory in Pasadena.

The planet apparently has gone through little change since it was formed. Its composition now should show what the solar system was like at its beginning and perhaps provide more data to substantiate the Big Bang theory.

Exploring the solar system and beyond will not be restricted to spacecraft probes. With the launching of the Hubble Telescope, we will, for the first time, be able to see much further into space. Until now our range of vision has been limited by the earth's atmosphere, which causes great distortions. Once in place 330 miles above Earth, the Hubble Space Telescope will see objects ten times clearer and be able to detect light that is 25 percent fainter than earthbound telescopes can now.

Andrew Fraknoi, executive director of the Astronomical Society of the Pacific, a nonprofit group in San Francisco, has said, "The space telescope will be to our time what Galileo's first telescope in 1609 was to his era. Galileo used his first telescope to discover the moons of Jupiter, the phases of Venus, and to make other fundamental discoveries. With the space telescope, equally dramatic discoveries are expected."

What the space telescope will see is open to wide speculation.

"The most exciting are likely to be objects we can't even imagine today," said Princeton University astronomer Lyman Spitzer, who has been called the "father of the Hubble Space Telescope." "Quasers were unheard of before the two-hundred-inch telescope at Palomar Observatory, near San Diego, and before Edwin Hubble's work with the one-hundred-inch on Mount Wilson, only sixty years ago, most of us thought the Milky Way was the whole universe. The Hubble Telescope represents a bigger leap in performance than either of these."

The telescope should be in orbit at the beginning of the nineties, although predicting exactly when is difficult, considering the project's history. The telescope has been sitting in storage—at a cost of $10 million a month—since November 1, 1984. It has been scheduled to be taken up by one of the space shuttles in October, 1986, but the Challenger disaster cancelled all flights for years.

As it turned out, it was for the best it did not get launched then, since the software designed to carry out the intricate maneuverings and positioning of the telescope in space proved greatly flawed. Supposedly, the flaws were worked out, and astronomers are clamoring to get man's first clear view of the universe.

While the space telescope will enhance our ability to see into the universe, other researchers are working on making contact with any intelligent life that may be out there.

It is not easy to reach out and touch someone in the cosmos. And if something out there is calling us, we are unaware of it. That is not to say, however, that we have not been trying to pick up any such signals.

It was in 1960 that a systematic attempt to pick up interstellar transmission began. Called Project Ozma (after the princess in Frank Baum's Oz series), it was the brainchild of Dr. Frank Drake. An 85-foot antenna at the National Radio Astronomy Observatory in Green Bank, West Virginia, was used to listen for any transmissions coming from two stars, located approximately eleven light years from Earth.

More recently, Project Sentinel, headed by Paul Horowitz of the Planetary Society and backed by Harvard University, used an advanced radio telescope at Oak Ridge Observatory to scan for transmissions along a wide band of 130,000 wavelengths. Horowitz predicts that the telescope's capability will increase to 8.4 million wavelengths.

NASA is also looking for intelligent life with its Search for Extra-terrestrial Intelligence (SETI) project. According to the space administration, "It is now possible to detect [radio] signals from anywhere in our Galaxy, opening up the study of over 100 billion candidate stars." In other words, if anything is out there, we are increasingly prepared to pick up the message.

The question still remains: How do we send our message out, beyond painting pictures on the Voyager spacecraft? We are ham-pered by the law of physics that no object travels faster than the speed of light. This is where the "neutrino" makes its entrance.

In the early 1930s, physicist Wolfgang Pauli suggested the existence of subatomic particles that could only be detected by their secondary or "residual" spin. Having no stable mass, neutrinos run into little resistance as they travel. Regular neutrons can only travel several feet in water before bumping into an atom of hydrogen or oxygen. Neutrinos would be able to travel for distances measured in light years.

With a neutrino beam, messages could be pulsed out through the galaxy with little or no interference from dust, stars, or nebulae. They would, therefore, have a greater likelihood of reaching distant receptive life than ordinary radio waves.

While contacting intelligent life would satisfy the innate curiosity of many people, for others the allure of the space program lies in dealing with pure science. For these people, finding clues and answers to questions that will advance our understanding of Earth, the solar system, and the universe is what the effort is all about.

The space commission, as part of its recommendations, urged that the space program be aimed at "understanding the evolutionary processes in the Universe that led to its present characteristics (including those leading to the emergence and survival of life)" and that we use this understanding "to forecast future phenomena quantitatively, particularly those that affect or are affected by human activity."

Some of the areas the commission had in mind were exploring the birth of the universe and the formation of galaxies, neutron stars, black holes, stars, and planets; determining how energy flows from within the sun outward to its surface and into space; learning about how the amount of energy from the sun varies and how the variation creates changes in the earth's atmosphere, causing such phenemona as Ice Ages.

From the vantage point of space, answers to questions closer to home will also be found, questions like: What makes up the earth's interior and crust? From where does the earth's magnetic field come? How are humans affecting the oceans, atmosphere, and crust of the earth?

As is often the case, these considerations will lead to practical applications. By the next century, we will have thirty-day weather forecasts that are 95 percent accurate. Earthquakes will be predictable twenty-four hours in advance, and hurricanes within twelve hours and one hundred miles.

Practical applications will be joined by intensified commercialization. The days of space being the exclusive bailiwick of the scientist, engineer, and explorer are over. The entrepreneur and the businessman will be joining them.

There has been a degree of commercialization already. Before the Challenger disaster, NASA was sending up commercial satellites. After Challenger, that task was left to Arianespace, the commercial satellite-launching arm of the European Space Agency. Ariane put in place numerous television, communications, and weather satellites for private concerns. In fact, there were so many satellites going up that Third World countries began to complain they were being crowded out of geostationary orbit—where satellites, traveling at the same pace as the rotation of the earth, stay in the same spot above the earth.

The Soviets, too, have gotten the jump on space commercialization. It began in September of 1989, when a three-stage booster rocket blasted off with advertisements emblazoned on its sides for an Italian insurance company, a Soviet electronics firm, and a Soviet perfume called "New Dawn."

Then, in February, 1990, two cosmonauts took off from the Baikonur space complex, located on the steppes of Central Asia, with the goal of making money in space. The six-month mission of Anatoly Solovyov and Alexander Balandin was to manufacture technological and biotechnological materials in zero-gravity conditions in a new module, Kirstall, that would be connected to the existing Mir orbital platform.

The Tass news agency estimated the mission would make up to $41 million in profits. "For the first time in Soviet cosmonautics, incomes from the flight are expected to exceed by far spending on the spaceship's launch."

Among Solovyov and Balandin's tasks will be making extremely pure semiconductor materials. Tass reported at the time of the launch that "Factories that placed their orders with the cosmonauts are waiting impatiently to put them into use."

But all this is a mere drop in the Big Dipper as entrepreneurs and corporations start seriously searching for ways to exploit space. Where will the profit opportunities of the future be found?

There is still money to be made in satellite communications. In the early nineties, cars, boats, and airplanes will be equipped with a receiver and display to pinpoint their exact locations. This will allow a driver to map out his route from driveway to destination. Pilots will receive collision warning, navy fleets will receive dispatch orders. Companies will be able to control remote factories by bouncing commands off the new satellites.

Satellites used for this remote sensing will find additional applications. Images of the earth produced by these satellites will allow better observation and management of crops, demographic patterns, pollution, and forests.

While communications satellites offer a sizable market of several billion dollars a year, the market for electricity is far larger, perhaps $400 billion worldwide. Improving space technologies offer the promise that at the beginning of the next century, satellites in geostationary orbit will capture solar energy for electrical use on earth.

It can also be expected that space factories will be built to manufacture new products. There will be new alloys formed from metals impossible to combine on earth because of gravity. Ultrapure crystals needed in the manufacture of semiconductors are likely to be produced in space factories.

Initially, because of the cost of transportation, microgravity manufacturing may be limited to products that do not weigh much and have a high retail price, such as drugs, electronic chips, and certain alloys. However, as the costs for getting materials to the plant and back to market diminish, the potential of what can be manufactured in space will rise.

And as the costs of transport fall and space travel becomes more routine, another industry will emerge. The Fun Industry. People dream of blasting off the mother planet. They want "to go, to touch, to see, to smell, to learn," as Neil Armstrong put it. Then they will want to return to Toledo and show their neighbors the videotape of the adventure.

In all likelihood, the early tours will be akin to the Cruises to Nowhere on the ocean liner *Queen Elizabeth II*. Go up, float around in zero-gravity, come back down. Then as more space stations are built, quick tours of them will be added to the itinerary. Finally, full-sized resort and recreational facilities will be built. Disney Universe might supplant Disney World. There might be a whole spin-off mini-industry in developing new sports for Club Milky Way.

Commercialization will follow the progress into space. As we establish ourselves on the moon and eventually Mars, outer-space industry will change focus. By as early as the mid-twenty-first century, space factories will no longer be producing products solely for use on earth. Space industries will have developed solely for markets in space. The moon will be one likely source of raw materials. Its soil will be used for radiation shielding, its oxygen in rocket propellants, its glasses in building, and its silicon in solar power panels.

As we continue space development with bases on the moon and Mars, we will have an increased need to cut the umbilical cord to earth. "Self-replicating" remote-control operated factories will be built. They will have transport machines, processing plants for raw materials, and shops for making parts for new plants. The raw materials found in space will be turned into finished product in outer space, saving the high transport costs to and from earth.

There are some who ask "Why?" Why all this effort and expenditure? After all, as the oceanic, cobalt-blue planet Earth rotates around its blazing orange-yellow sun, it is merely one of many shimmers of light in the Milky Way. That galaxy, for all its size and magnitude, shines insignificantly in the vastness of the dark universe. And yet, from the minutia that is earth echoes loudly the intuitive belief that space is the ultimate frontier for us to conquer.

Space exploration exists not only for utilitarian reasons, for raw material and breathing space, but for something noble—to inspire mankind. Colonization of the moon and the establishment of planetary settlements in our solar system and beyond should inflame rather than satisfy our primeval urge to explore.

Though it is possible to search on for philosophical motivations or genetic imperatives, it may have been a nine-year-old who best answered the question of why man will continue to venture away from Earth. "Space," she said, "is neat."

TEN

Getting Ready for
the Millennium

RECOGNIZING THAT CHANCE, in all its capriciousness, deals many hands, man makes much of his own fortune. This book has endeavored to explore the technologies and trends, to find the breakthroughs and advances that will be our future. But merely knowing what to expect is not enough. We must be prepared to use and exploit to the fullest what tomorrow has to offer. We must be ready to meet the new millennium.

Millennium has more than one definition. According to Webster's New World Dictionary, it means "a period of a thousand years" and "any period of great happiness, peace, prosperity, etc.; imagined golden age."

We have the potential of at least partially attaining that second definition. Our next thousand years could be a golden age, but we as a nation and we as a world must move in new directions without being tied to the errors of the past.

For the U.S. best to exploit what is on the horizon, it must ready itself for the future now. We have to make adjustments. Third World countries often receive trickle-down technology. They lag behind because they are in no position to initiate. They often suffer economic and social impotence. The United States could fall into the same position.

It does not take a degree from the Wharton School of Business to know that the American economy has been in a precarious state. When polled in the mid-eighties, almost half of 105 chief executive officers of Fortune 500 companies saw a major financial crisis looming before 1995. Such pessimism should come as no surprise. Symptoms of economic malaise have been manifesting themselves for years, some insidiously, others with the force of a Mike Tyson punch. As the national debt accelerated, real economic growth slowed. In the sixties that growth rate was 4 percent, in the seventies 3 percent. And in the eighties? The real growth rate was an almost nonexistent 2 percent. At the same time total debt—private and public—skyrocketed.

Such debt could have fueled growth if the capital had been used for new fixed investment. It was not. The inflow of capital was spent on interest payments, government securities, and current consumption.

Because of this lack of new capital investment, the economy is operating below its potential. Economists at Morgan Stanley, the New York investment banking house, estimated the national output was approximately 5 percent lower than it could be. To put that into perspective, at the end of the 1980s, the United States had a $4 trillion economy. A 5 percent increase would mean an additional $200 billion—the equivalent of the yearly national budget deficit.

Obviously, something more than Chrysler chairman Lee Iacocca's exhorting college students to get out and make a difference is needed. The United States must face up to the fiscal irresponsibility of the past forty years and be prepared to take some strong medicine. A wide spectrum of policies needs to be instituted if the United States' economic system is to regain stability. Of course, no complete list of policies can be written in stone today. Circumstance and chance will necessitate further adjustments. Acknowledging the unforeseen, a blueprint for the future still may be rendered now. At the least, the following policies should be included in that plan.

1. Freeze federal spending to save $50 billion annually. Establish a ceiling on federal, state, and local government spending based on programmed growth. That is, any increase would be tied to population growth and new service demand. The increase would also be tied to inflation and would not be more than 50 percent of the previous year's inflation rate. California already has this in place in the guise of Proposition 4.

2. Cut back on payments to NATO. At the end of the eighties, each person in the United States was contributing $669 a year to NATO, sixty percent of all military spending. Lower our contribution to only $20 billion.

Hoover Institute fellow Melvyn Krauss makes the cogent argument that the U.S. has in essence spoiled our European allies. We pick up the major portion of the defense tab. We provide the nuclear umbrella. The United States also wields most of the power. So, on the one hand, such a policy does little to motivate the Europeans to provide their own defense, while on the other, it creates anti-American feelings because the Pentagon is executing the major decisions. Krauss goes even further than payment reduction. He would like to see all U.S. troops out of Western Europe. Should the changes in Eastern Europe and the Soviet Union continue at their present pace, the need for NATO will decrease even further.

3. Start selling off the nearly $300 billion in notes the federal government holds on, among other things, foreign nations, corporations, student loans, farms, and mortgages. By wholesaling these notes, the government could raise in the neighborhood of $150 billion, which it could immediately apply to the national debt. There is no reason the loan portfolio of the United States Treasury should be larger than the portfolios of the country's two largest commercial banks combined.

4. Implement a federal auction program to solve the nation's savings and loan crisis. Instead of saddling American taxpayers with additional taxes totalling more than $157.6 billion (Treasury Secretary Nicholas F. Brady's estimate of the cost, including interest, of the Bush administration's bailout plan), the federal government

should auction the properties and securities of insolvent savings and loan institutions to generate as much as $120 billion in proceeds. These funds could be earmarked to aid troubled, but still solvent, savings and loan institutions. Buyers' premiums of 10 percent of the total auction proceeds would account for additional income to the federal government.

5.   The federal minimum wage should be tied to the inflation rate and come under prescribed periodic review. Raising the minimum wage has too often become a political football at the expense of a segment of the American population that is highly vulnerable.

6.   Reorganize the Pentagon's Procurement Systems Program.

There is something shamefully wrong when our military's buying habits are standard jokes on the Johnny Carson Show. If John Doe can walk into a K-Mart and buy a toilet seat for twenty dollars, the Pentagon should not be paying several hundreds of dollars for a similar seat. The country cannot continue to absorb this magnitude of waste.

To streamline and bring sanity into the acquisitions process, a professional civilian procurement group should be established. It should be recognized, of course, that buying for the Pentagon is not the same as meandering into a couple of stores and comparing features and prices on lawn mowers. A corps of highly trained people is needed to handle government procurement.

For this new acquisitions process to operate more efficiently, the Pentagon's budget cycle should go from one year to two years. The Presidential Commission on Pentagon Management has recommended the institution of five-year spending plans. Too much time and effort is wasted haggling over the budget every year. As it is, there are thirty-two congressional committees, each with a staff, involved in the military procurement process. The military-congressional complex needs to be reduced to a manageable size.

Another gargantuan problem has to be faced—cost overruns. Take the Phoenix, the air-to-air missile being built for the Pentagon by Hughes Aircraft. According to a Pentagon inspector general's report released at the end of 1988, it was costing three times as much to build the missile as the contractor had originally estimated.

Hughes had said that operating at peak efficiency (the Pentagon should have seen a giant red flag there with flares going off as well), it would take 190,000 hours to build 265 missiles. The Navy ended up paying for 713,000 hours. Why? Inefficiency.

Another Pentagon study found that contractors normally went 87 percent over time estimates. Reasons or excuses—depending on your viewpoint and magnanimity—are offered for each case of overrun, but one underlying cause can usually be found among all the other reasons. The contractors' plants are in need of modernization and new production technologies to keep costs and overruns down, to make them more efficient. Their plants and systems are not up to meeting the demands of the multiplicity of military projects. Modernization takes money and an eye for potential future profits. Most companies, worried with this year's bottom line or at most a few years beyond that, are unwilling to make the investment. Strategic Japanese firms generally evaluate projects in time frames of more than ten years.

One solution to this dilemma is something called Get PRICE—Get Productivity Realized through Incentivizing Contractor Efficiency (an acronym only the military could love). The brainchild of Air Force Lieutenant General James Stansberry, this program held out a carrot to contractors. The Air Force would share with the contractor any savings realized from having a project done in a modern, state-of-the-art plant.

It is not a bad deal for the contractor. It gets to try out new technologies—which, if successful, can then be sold. New plants are financed which can make it more competitive in nonmilitary projects as well. It also builds up goodwill with the Pentagon for future projects. As Thomas Murrin, president of Westinghouse Electric's Energy & Advanced Technology group, an early participant in the Get PRICE strategy, said, "Get PRICE fosters a spirit of cooperation between business and government rather than antagonism. I would much prefer the carrot to the club."

7.   Grant the president line-item veto on all budgetary matters.

Our legislative process, as it is now arranged, is well-liked by congressmen in need of brownie points back home. One merely needs to squirrel in a budget item that will make the constituents happy. If that particular expenditure meets with the President's disapproval, he has to veto the entire bill to reject it. While it may be argued that is part of the checks and balances between the executive and legislative branches of government, in reality a lot of unneeded checks get written to keep this balance.

Congressmen play intricate games among themselves even before a bill reaches the President's desk. "Let my pork barrel roll, and I won't stand in the way of yours."

If the President is allowed to deal with each expenditure separately, then it will be up to Congress to override the veto.

Forty-three states have already instituted line-item vetoes. "Governors find this a weapon in maintaining financial discipline," according to M. S. Forbes, Jr., deputy editor-in-chief of *Forbes* magazine.

8.   Totally overhaul the Social Security System along the lines of the plan suggested by Stanford University's Michael Boskin.

If Social Security is left as it is, there could well be intergenerational warfare in the future as younger workers are asked to pay more and more into the system to provide benefits for gray-haired baby boomers.

The Boskin plan calls for separating insurance benefits from transfer-payment functions, basing payments on individual needs, and linking total benefits received to lifetime contributions plus a normal-rate-of-return.

Along with Social Security, the federal pension system needs reform. As it stands, military personnel receive 75 percent of their preretirement salaries. In the private sector, retirees can expect to receive only 50 percent.

9.   Funnel subsidies away from the farms that don't need them and to the ones that do. In 1986, more than 100,000 farmers did not have money for spring planting. At the same time the largest farms in the country—about 4 percent of all farms—received more than

fifty percent of the federal subsidies. Redirect some of the capital that has been going to the megafarms into a National Save the Farm Program.

Further, a fund for rural development, as supported by Senator Robert Dole, could provide loan guarantees for farmers in need of financing.

10.   Fund a National Redevelopment Bank along the lines of the Industrial Bank of Japan. In the United States, when companies have severe financial setbacks, bankers doublecheck that all loans are secured so they will be the first to get paid when the company files for bankruptcy. In Japan, the banks step in to assist the corporation. The Industrial Bank of Japan has been a major player in these restructurings, from the early post-war period when Nissan Motors was foundering, into the sixties with Yamaichi Securities, and more recently with Chisso Chemical.

It is in the best interest of the banks for a company to be solvent, even if the banks do have their loans repaid. A defunct company is a lost customer.

11.   Bring back investment credits for troubled industries. The credit would stimulate modernization of plants and equipment. Harvard economist Lawrence H. Summers was one who argued against the repeal of the investment tax credit. "The effect is to lower taxes on old investment and increase them on new investment. That's exactly backwards."

12.   Shift to differential depreciations rates on an industry to industry basis. The rates would be set within the specific industry based on comparative five-year-growth rates. In this way, heavy capital-intensive industries that have long been in a state of decline would have a decreased tax burden.

13.   Give special tax benefits to small businesses, which represent 80 percent of all United States business operations. Let small businesses deduct 100 percent of their health insurance premiums. Raise the amount of new equipment subject to immediate write-off from $5,000 to $50,000. Create venture-capital debentures that would be convertible into common stock. Give special capital

gains status to this debenture, encouraging investments in new companies.

14.   To combat the dearth of young people entering certain key professions, institute a ten-year tax deduction to entice more students into these crucial fields. A good example would be engineering. Less than 10 percent of American undergraduate degrees are in engineering. This compares unfavorably with the 35 percent in the Soviet Union and 37 percent in West Germany. If the U.S. is to avoid a debilitating technogap in the next century, more emphasis must be placed on educational opportunities in engineering.

15.   Start a national lottery.

Not too long ago there was something immoral, decadent, Shanghai-in-the-twenties about government's getting into the gaming business. Admittedly, it was legal in Las Vegas, but somehow that city had an unsavory connotation. Times have changed. Off-track betting came to New York, and lottery after lottery came to state after state. It is time for the federal government to go along with the trend.

Money raised from the lottery could be used in a national reemployment program. Only about $160 million is now being funneled from Washington to the states under the Federal Job Training Partnership Act for job training programs. Using lottery money, state programs could conceivably receive three times that much.

16.   Approve a national initiative so that any issue can be put on the ballot if enough signatures are collected within an eighteen-month period. If passed, the initiative would then be executed into law. John Naisbitt, author of *Megatrends,* argues that initiatives have proved effective on state and local levels where already "citizens are deciding by popular vote issues that only they will have to live by."

17.   Abolish the Electoral College. It is an anachronism, or as former House Majority Leader Jim Wright once put it, "a relic of the powdered wig and snuffbox era." True, there has been only one presidential election, that of 1888 between Harrison and Cleveland, where a candidate with more popular votes lost because his

opponent had the edge in the Electoral College. (In 1876, Tilden had more popular votes than Hayes and also lost, but that was because the returns in Florida, Louisiana, Oregon, and South Carolina were disputed and a joint session of Congress declared Hayes the winner.) Because the Electoral College overrode the popular vote only once in history does not mean it could not happen again. Doing away with the Electoral College, through a constitutional amendment, would assure the tenet of One Man, or rather One Person, One Vote.

18.   Curb the power of Political Action Committees (PACs) by increasing the campaign income-tax checkoff from $1 to $3. The checkoff money could then be used not only in presidential races, but congressional as well.

19.   Bring back the complete Individual Retirement Account (IRA) tax deduction. Before the tax changes, anyone, whether he or she had another pension plan, could set up an IRA with tax-deferred earnings. The theory being that the tax would be paid after retirement at a lower rate. While the IRA deductions were costing the Treasury about $6 billion a year, the benefits far outweighed the loss.

20.   Pass parental leave and child care legislation. The cost of benefits to the millions of working-out-of-the-home mothers in the 1990s will be far outweighed by greater stability in the workforce.

21.   Place a $5 import fee on every barrel of oil brought into the United States. Once OPEC oil prices dropped, there was little incentive to continue domestic oil exploration and production. Conservation efforts fell by the wayside. Interest in developing alternative energy sources dropped because those sources could not compete in price with cheaper oil.

However, there is no reason to believe oil prices will always be so low. Nor, for security reasons, is it sound policy to depend so heavily on foreign imports. Make foreign oil more expensive, and domestic oil and alternative energy sources would once again look more desirable.

The additional money brought in by the fee could be used in any number of ways. It could go toward reducing the national debt if it could be used more imaginatively. For instance, it might go toward environmental clean-up programs. We now have almost nine hundred Superfund sites, and the list is expected to more than double in the next ten years. Spend the oil-import revenues on scrubbing that list clean or on finding alternative energy sources.

22. Establish a comprehensive public policy in support of research and commercialization of alternative energy technology. Make the fuel-economy standards more stringent. Raise the gasoline tax and use the additional revenue to finance new mass transit systems. It is estimated that every full train car takes 75 to 125 cars off the road.

23. Create a National Export Bureau to act as a liaison between multinational U.S. corporations and foreign buyers. The bureau's main purpose would be to stimulate foreign business for U.S. corporations. It would provide product samples, literature, price lists, and formal introductions. Ireland has a similar export agency and has seen a two-digit increase in export revenues over a five-year period.

On an international level, no golden millennium will ever be reached without monetary reform. Parliaments, congresses, and their respective national governments must be forced to understand the magnitude of the international debt crisis and the inherent dangers in our current system. Armed with these insights, our leaders must then summon the political courage to carry out at least some of the following measures.

1. Establish exchange rates that would be adjusted variably to prices on a ninety-day basis. Such adjustments would make corrections for inflation and would tend to have an impact over longer periods. (Depreciation of the dollar in 1977–78 generated a trade surplus two years later, consistent with the so-called J curve theory of lagged variables.)

2. Create a single global monetary medium of exchange, the International Unit of Currency (IUC), that would be fully convertible into all currencies and redeemable in gold. IUCs should be

transferable on CHIPS, the Clearing House-Interbank Payments System. In the 1990s, CHIPS will electronically transfer more than $500 billion daily among world banks, compared to approximately $300 billion in 1986. Since investment demands are virtually swamping transaction demands for the yen and dollar, creation of the IUC would improve capital market efficiency.

3.   Encourage development programs in the Third World. Grant business development holding companies the right to collateralize development projects in Third World nations.

4.   Establish a special division of the International Monetary Fund to restructure outstanding foreign debt so that surpluses could be more effectively recycled to the Third World. Unfortunately, net private loans to "problem" nations have been declining. The new IMF division should be allowed to purchase loans at net discounts on the face value and pass these "debt retirement savings" on to the borrowers.

5.   Implement conversion of Third World debt into securities backed by future production of petroleum, natural gas, minerals (including gold, silver, copper, and uranium), and food. Conversion through natural resource collateralization would dramatically increase the monetary capital bases of numerous Third World nations. Allow Third World nations to exchange stock backed by natural resources for existing loan obligations.

6.   Allow large debtor nations to finance substantial portions of their debt through commercial banks on a fully secured, collateralized basis at lower interest rates. Nations such as Egypt, which had debts totalling more than $26 billion in 1986, could save hundreds of millions of dollars annually by "repositioning" their debt.

7.   Set up repayment schedules for Third World countries based on a percentage of their export earnings. On average, debtor nations can channel as much as 25 percent of their yearly export income to debt repayment without limiting their economic growth.

8.   Force nations that cannot make payments on debts with export income to adopt a policy of matching spending with receipts. After Bolivia adopted such a policy in 1985, its inflation rate

dropped dramatically in only one year.

9. Institute automatic wage and price freezes when a debtor nation's inflation rate surges past 100 percent. When Brazil announced wage and price controls after its inflation rate neared 400 percent, that policy, combined with a termination of indexing, led to a sharp decline in inflation as well as an $11 billion trade surplus the subsequent year.

10. Shift toward more decentralized "free market development" of Third World countries. This would mean opening domestic markets to free trade, setting up free-enterprise zones, and encouraging foreign investment through tax incentives.

11. Support Third World nations that sell off state-owned enterprises or convert such concerns into publicly held companies and then use the proceeds to retire debts.

12. Stop lending money to corrupt state-owned enterprises. At one time in Ghana, 90 percent of all foreign loans was going to government-controlled concerns that never reallocated the funds to the private sector.

13. Shift development aid in Third World countries from large-scale state-owned projects to specific decentralized ones. For example, instead of spending $800 million to build two dams on the Senegal River in Africa, distribute the funds among the 400,000 villages of the sub-Saharan Sahel Region so that water wells might be dug and small photovoltaic solar electricity plants built.

14. Allocate a portion of development aid to vocational training programs. Creation of a semiskilled work force will make emerging nations desirable locations for multinational corporations.

These reforms should be executed with even greater constructive changes if we are to attain our golden millennium. There must be a transformation of society in the next century that will include a redistribution of the world's capital base, totally new patterns of transnational development, minorities having a more equitable share of power, and wide development of alternative energy.

So that society may enjoy greater personal freedoms, a fairer distribution of wealth, and a higher semblance of peace, we will have to eschew short-term profit incentives in favor of altruistic long-range goals. In the words of James M. Buchanan, a Nobel laureate in economics, "It is time to start replacing dystopia with a tempered utopia."

In our free society, the enduring belief that one person can make a difference still holds true. And, even if it has done nothing else, the new Information Age has increased the American sensitivity to social injustice. "The real news in American society," wrote Walter B. Wriston in *Risk and Other Four-Letter Words,* is "not that our problems are great and have multipled but rather that our sensitivity to these problems is greater now than at any time in the long history of man."

We have channeled our increased awareness of wrongs and injustices into public interest causes. More Americans are getting involved in championing arms control, civil rights, protection of the environment, public health, welfare, and education, as well as pushing for more enlightened economic regulation.

In coming years we will be turning our energies toward keeping new technology manageable and under control. Our goal will be to provide clarity and conscience to future scientific breakthroughs and their applications.

Charles, the Prince of Wales, has recognized the danger of mankind being overwhelmed by technology. In an address at Harvard University's 350th anniversary, he said, "To avert disaster, we have to not only teach men to make things, but to teach them to have complete moral control over the things they make . . . it is possible then to see a relationship between moral values and the uses of science."

Undoubtedly our greatest misallocation of technology has been in the billions of dollars spent on advanced military systems. There are so many more worthwhile uses for that money. Masaki Nakajima, former chairman of Japan's Mitsubishi Research Institute Incorporated, once called for a global fund that would, among other things, provide transportation and communication links to Third

World countries and pay for new food production systems, water and power projects there. All of this would have cost just half of the world's yearly defense budget.

In *Arms and Hunger,* former German Chancellor Willy Brandt compared the costs of selected military and civilian programs. He found that 160 million children in 23 developing nations could be educated for what it costs to build one nuclear submarine; forty thousand village pharmacies could be kept stocked for the price of one fighter bomber; and one thousand classrooms accommodating thirty thousand children could be set up for what it costs to build a single tank.

Current attempts at arms reduction, however limited in scope, should be wholeheartedly encouraged while at the same time further cuts should be demanded by citizens on both sides of the crumbling Iron Curtain.

It is time we took mankind off the endangered species list by becoming free of nuclear weapons by the year 2000.

We should not stop on earth. The 1967 Outer Space Treaty should be reaffirmed. No country should send nuclear weapons into earth orbit, nor should they be placed in outer space. Establishment of military bases and the testing of weapons on other planets should be prohibited. Man's "final frontier" of space should be forever a demilitarized zone for peaceful international cooperation. We have far too many problems on earth to create additional ones in outer space.

In the coming years we must address the issues of eradicating world hunger, banning biological weapons, and setting guidelines for genetic engineering research. We must also come to terms with the increasing disenchantment with our national governments as vehicles for solving mankind's dilemmas. Contemporary life is marked by the utter failure of our ruling ideologies to deliver a desirable and decent life to the citizenry.

The future of civilization depends on replacing the hopelessness of unresponsive economic and political systems with dynamic, strategic policies. The human race is now engaged in a great evolutionary leap forward. Enlightened policies of the future will spell the difference between avoiding a new Dark Age and the creation of a glorious new millennium which has the capacity of transforming mankind forever.

Index